MRE

Materials Research and Engineering
Edited by B. Ilschner and N. J. Grant

Maciej Pietrzyk · John G. Lenard

Thermal-Mechanical Modelling of the Flat Rolling Process

With 97 Figures

Springer-Verlag
Berlin Heidelberg NewYork
London Paris Tokyo
Hong Kong Barcelona

Dr. MACIEJ PIETRZYK
Department of Metallurgical Engineering
Mining and Metallurgy University
Mickiewicza 30, 30-059 Krakow, Poland

Prof. JOHN G. LENARD
Department of Mechanical Engineering
University of Waterloo
Waterloo, Ontario, Canada, N2L 3G1

Series Editors

Prof. BERNHARD ILSCHNER
MCX-D Ecublens Ecole Polytechnique Fédérale de Lausanne
Laboratoire de Métallurgie Mécanique
CH-1015 Lausanne/Switzerland

Prof. NICHOLAS J. GRANT
Department of Materials Science and Engineering
Massachusetts Institute of Technology
Cambridge, MA 02139/USA

ISBN 3-540-53316-8 Springer-Verlag Berlin Heidelberg NewYork
ISBN 0-387-53316-8 Springer-Verlag NewYork Berlin Heidelberg

Library of Congress Cataloging-in-Publication Data
Pietrzyk, Maciej
Thermal-mechanical modelling of the flat rolling process/Maciej Pietrzyk, John G. Lenard.
(Materials research and engineering)
ISBN 0-387-53316-8.-- ISBN 3-540-53316-8
1. Rolling (Metal-work)--Mathematical models.
I. Lenard, John G. II. Title. III. Series: Materials research and engineering (Unnumbered)
TS340.P47 1990
671.3'2--dc20 90-22568

Printed in Germany

Offsetprinting: Color-Druck Dorfi GmbH, Berlin; Bookbinding: Lüderitz & Bauer, Berlin
2161/3020-543210 - Printed on acid-free paper

PREFACE

The abundance of mathematical models describing the flat rolling process is indicative of its major technological importance. Indeed it is of some interest to note that the USA and Canada produced a total of 43 million tons of sheet and strip products in 1985, indicating that the steel industry's contribution to the gross national products of both countries is significant. In that context improvements of the models are of importance as is the answer to the question: of all models that are available, which is the most adequate for a particular application? A proper answer should be based on a comparative evaluation of the predictive capabilities of the models. As well, the sensitivity of the models to variations in parameters and to changes in initial assumptions should also be understood. The present book is an attempt to deal with these aspects of mathematical modelling.

A compilation, analysis and discussion of the authors' research concerning hot and cold flat rolling are presented in this book. The central theme is mathematical modelling of the processes and the main emphasis of the general approach is one of substantiation. In each case presented the predictions of the models are compared to measurements, obtained in both laboratory and industrial conditions. The models are evaluated in terms of their usefulness for on-line adaptive control.

The book contains six chapters. In the first, the Introduction, the writers' understanding of the term "mathematical modelling" is described and a general formulation of the flat rolling problem is presented. Conventional models of strip and plate rolling are given in Chapter 2, including those of Orowan and von Karman and their simplified versions such as Sims' model for hot rolling and Bland and Ford's for cold rolling. Tselikov's technique, applicable for either hot or cold rolling, is then described. The empirical formulae of Ford and Alexander are also listed. After a critical comparison of the assumptions of the models and a statistical evaluation of their predictive capabilities, a refinement, which addresses criticisms levelled against the Orowan technique, is discussed in Chapter 3.

The thermal events occurring during both hot and cold rolling are considered next and are coupled with the mechanical phenomena in a complete thermal-mechanical finite-element model of flat rolling which is presented in Chapter 4. The finite element approach is used for both components of the model. The elastic deformation of the roll is accounted for. The metals' resistance to deformation is expressed in terms of strains, strain rates, temperatures and in the case of carbon steels, in terms of the carbon content, as well. The model's predictions of the temperature fields are compared to experimental results obtained in a laboratory setting during cold, warm and hot rolling of aluminum and steel strips and slabs. As well, comparison to results obtained on a seven-stand industrial strip mill is given.

Conventional modelling of the flat rolling process necessitates an assumption of a uniform stress distribution across the strip. This assumption becomes questionable for rolling of thick samples when the thickness is often much larger than the length of the deformation zone. This aspect of the rolling process is the topic of Chapter 5 which contains an account of the role of the shape coefficient in the modelling process.

Conclusions and discussion of the use of the models are included in the last chapter of the book. The general idea of modelling is considered, especially in the light of ever increasing use of computer-aided engineering. A preliminary answer to the question "Which model should be used in a particular set of circumstances?" is given. It is pointed out that in light of further research, advances in computing and computer technology and the potential availability of more reliable data on boundary conditions, that answer may well have to be reformulated.

The authors are grateful for the support received from various agencies and organizations. These include the Natural Sciences and Engineering Research Council of Canada, the American Iron and Steel Institute, NATO, Department of Supply and Services of Canada, CANMET, Department of Energy, Mines and Resources of Canada, Bethlehem Steel Corporation, Republic Steel Corporation, Algoma Steel Corporation Ltd., The Steel Company of Canada and Sidbec-Dosco Co. Ltd. The support of the Mining and Metallurgy University of Krakow, Poland, the University of New Brunswick and the University of Waterloo is gratefully acknowledged. As well, our colleagues and students who were a constant source of inspiration have to be named. For John G. Lenard, R. Roychoudhuri and A. Karagiozis should be mentioned. For Maciej Pietrzyk, M. Glowacki and J. Kusiak deserve special thanks. Finally, the place of honour must go to our families without whose constant support none of what follows would have been possible. Sincere gratitude is expressed to Harriet, Alina, Patti, Wojtek and Marta.

Further thanks are due to the publishers permitting reproductions of portions of our previous publications. In this context the authors are thankful to the Finnish Society for Tribology, the Editor of Heat and Technology, Ellis Horwood Limited, Steel Research, Kluwer Academic Publishers, The Japan Society for Technology of Plasticity, Elsevier Science Publishing Company Inc. and Energy, Mines and Resources Canada.

February, 1991

Maciej Pietrzyk
Mining and Metallurgy University
Krakow, Poland

John G. Lenard
University of Waterloo
Waterloo, Canada

CONTENTS

LIST OF SYMBOLS

b - width of the strip
$\mathbf{b}$ - vector of the body forces
c_p - specific heat
D - activation energy of diffusion
E - Young's modulus
E - identity matrix
F - roll separating force per width unit
$\underline{\mathrm{F}}$ - vector of tractions on the boundary
F_{e1}, F_{e2} - contributions of elastic entry and elastic exit to the roll force
f - plastic potential
h - current strip thickness
h_1, h_2 - initial and final thickness of the strip, respectively
h_{1e}, h_{2e} - entry end exit thickness for the plastic zone
h_n - strip thickness at the neutral point
k - yield stress in shear; conductivity; Boltzman constant
k_1, k_2 - yield stress in shear before and after the pass, respectively
l_3 - length of the dead zone
l_b - length of the backward slip zone
l_d - length of contact with a rigid roll
l_d' - length of the horizontal projection of the arc of contact
l_n - length of the forward slip zone
M - roll torque per unit width

$N, \hat{N}$ - shape function, surface shape function
$\mathbf{N}$ - matrix of the shape functions
n - rotational speed of the roll
$\mathbf{n}$ - vector normal to the surface
Pe - Peclet number
p - roll pressure
p_1, p_2 - roll pressure at entry and exit planes, respectively
p_{p2} - roll pressure at entry to the plastic zone
p_{p3} - roll pressure at exit from the plastic zone
Q - rate of the heat generation due to plastic work
q - rate of the heat generation due to friction forces
R - roll radius
R' - radius of curvature of the flattened roll in the deformation zone
r - reduction; coordinate in the cylindrical system
S - contact surface
s - forward slip
T - temperature
$\mathbf{T}$ - vector of nodal temperatures
T_1, T_2 - back and front tension per unit width, respectively
T_0 - roll surface temperature or ambient temperature
t - time
V - control volume
v - velocity
$\mathbf{v}$ - vector of nodal velocities
$\underline{\mathbf{v}}$ - vector of velocities on the contact surface
v_r - roll velocity
v_s - slip velocity
v_x, v_y - velocity components
v_1, v_2 - entry and exit velocity of the strip, respectively
$\dot{W}$ - power of the rolling mill
$\dot{W}_c$ - power dissipated due to velocity discontinuities

$\dot{W}_f$ - power for friction losses
$\dot{W}_p$ - power for plastic deformation
w - weighting function
x, y, z - coordinates
α - heat transfer coefficient
γ - dislocation density; angle between the tangent to the roll surface and the horizontal axis
Δ - shape coefficient
$\Delta v_x, \Delta v_y$ - increments of velocities
$\Delta \mathbf{v}$ - vector of increments of velocities
ϵ - true strain
ϵ_i - effective strain
ϵ_p - peak strain in compression test
$\dot{\epsilon}_i$ - effective strain rate
$\dot{\epsilon}_{ij}$ - components of the strain rate tensor
$\dot{\epsilon}_x, \dot{\epsilon}_y, \dot{\epsilon}_{xy}$ - strain rate components
$\dot{\epsilon}_v$ - volumetric strain rate
Φ - stress function; redundant strain factor
ϕ - coordinate in a cylindrical system
ϕ_n - angular coordinate of the neutral point
ϕ_1 - angle of bite
ξ - acceleration coefficient
λ - Lagrange multiplier
μ - coefficient of friction
μ_0 - constant value of the coefficient of friction
ν - energy per unit length of dislocation; Poisson's ratio
ψ - lever arm coefficient
ρ - density
ρ_0, ρ_1 - norms
σ - flow strength

σ_i - effective stress

σ_{ij} - components of the deviator stress tensor

σ_p - flow strength, as applied to modelling of the rolling process

σ_{p0} - initial yield stress of the material, immediately after annealing

$\sigma_x, \sigma_y, \sigma_{xy}$ - stress components

σ_1, σ_2 - back tension stress and front tension stress, respectively

σ_{e1}, σ_{e2} - back tension stress and front tension stress as affected by elastic entry and exit

τ - friction stress

τ_x, τ_y - components of the friction stress

ω - roll angular velocity

CHAPTER 1

INTRODUCTION

Computer control of strip and plate rolling equipment requires the use of a predictive -adaptive mathematical model of the process and of the equipment. The purpose of the predictive portion is to compute the process parameters, given the desired mechanical and metallurgical properties and the final dimensions of the product along with its initial metallurgical make-up. The adaptive part should be designed to make use of the results of the predictive model, analyze them in combination with data supplied by the sensors and monitor and correct the settings of the mill parameters during rolling. Further, it should have a learning component, storing all relevant information during production runs, for future use.

An excellent book has been published recently (Boër et al., 1986) dealing with modelling of metal forming processes. The authors discuss forging, roll forming, drawing and thermomechanical treatment in detail. Rigourous accounting for mechanical and thermal events during forming differentiates this book from others that analyze stresses and strain or temperatures separately.

This approach has been followed in the present detailed study of the flat rolling process. The central concern is the predictive component of the model. The term "mathematical model" is used here to indicate a set of equations, derived from first principles, that transforms a physical event into those equations. First principles include laws of nature, such as Newton's laws of equilibrium and action/reaction; material properties, measured in independent tests and expressed in terms of empirical relations, and boundary/initial conditions, expressing the state of stresses, strains, deformations, temperatures and their derivatives at particular times and locations.

Mathematical models of the flat rolling process are numerous. In each, the equations of motion, thermal balance, material properties and roll deformation are used to calculate the stress, strain, strain rate, velocity and temperature fields, the roll pressure distribution, roll separating forces and roll torques.

The consistency and accuracy of the predictive capabilities of these methods of computation depend exclusively on the quality of assumptions made either during the derivation or the solution of the basic equations. In the conventional models, researchers, almost without exception assume the existence of homogeneous compression of the strip, considered to be made of an isotropic and homogeneous material that is incompressible in the plastic state. Further, plane strain conditions are assumed to exist and either a constant friction factor or Coulomb friction conditions apply at the roll-strip interface. Assumptions and simplifications vary broadly when finite element methods are employed. Two and three dimensional models have been presented, treating rigid-plastic or elastic-plastic materials; roll elastic flattening is sometimes included but often ignored; frictional conditions are described in a variety of ways. Excellent reviews of the theory, analysis and practice of flat rolling are presented by Roberts (1978, 1983, 1988).

The success of a mathematical model is measured by its predictive ability. To establish that, a large number of measurements should be taken over a broad range of the experimental variables and the same variables should be calculated i.e. predicted by the model. The average percentage difference of predictions and measurements will indicate the accuracy of the model and the standard deviation of those differences will indicate its consistency. It is the latter that is of importance because a sometimes accurate predictive model is useless. However, an inaccurate but consistent model can always be used with appropriate tuning parameters.

As far as rolling is concerned, a complete mathematical model should include:

(a) equations of motion of the deforming metal,
(b) heat balance of the roll/strip system,
(c) equations of equilibrium of the work roll,
(d) description of the friction between the work roll and the metal,
(e) description of material properties.

A general mathematical model of the flat rolling process may be formulated following the discussion above. As the strip enters the roll gap it is first deformed elastically. It speeds up; the relative velocity between the roll and the strip is such that friction draws the metal in. The criterion of plastic flow governs the manner in which the transformation from elastic to plastic states happens. As Fig.1.1 shows, the two states are separated by a surface called the elastic-plastic interface. The strip proceeds through the roll gap and more plastic flow occurs until finally at exit the roll pressure is removed. The strip is unloaded and it returns, through an elastic state to the original, load free condition. It is observed that during rolling, the relative velocities of the roll and the strip change and as the strip is accelerating forward it reaches the roll surface velocity at the no-slip or neutral point. From then on, as further compression occurs, the strip speeds up and the direction of friction changes in such a way that it now retards motion. Exit velocity of the strip is often larger than that of the roll and the difference between the two velocities is determined by the forward slip.

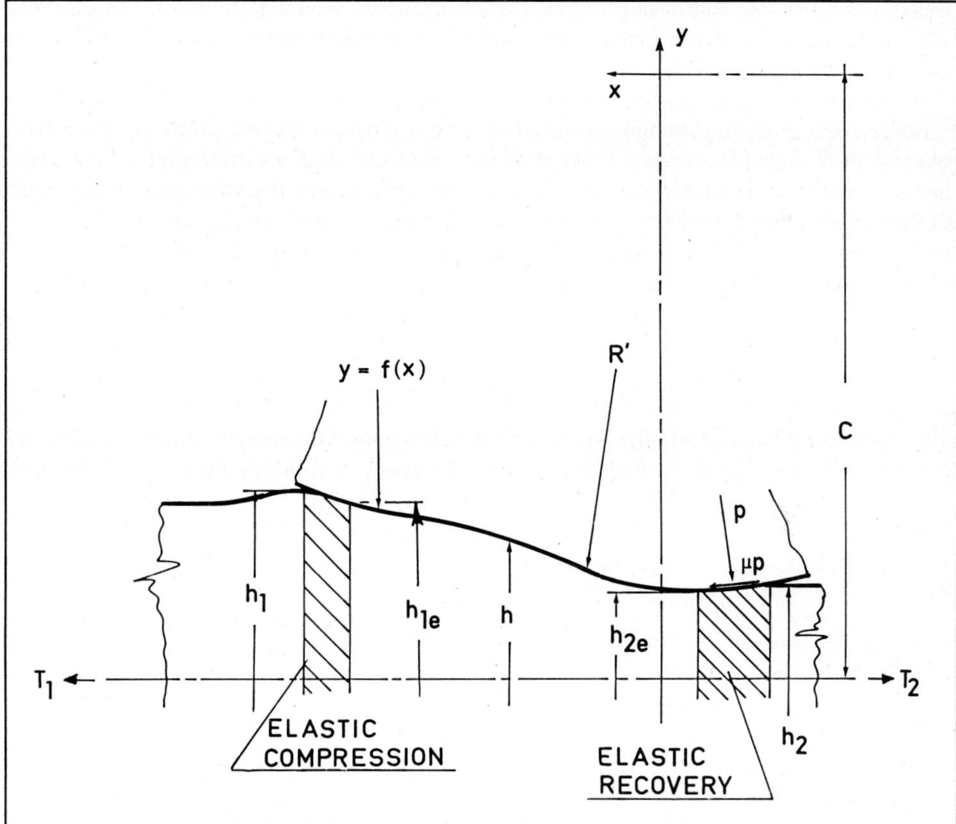

Fig.1.1 Schematic diagram of the flat rolling process.

Thermal events, occurring during the strip's passage through the roll gap are also of importance. In fact, it is these that contribute most to the metallurgical development of the final structure of the rolled material. Surface conditions, roll wear and thus roll life are also affected by thermal conditions. Heat is generated because of the work done on the strip, increasing its temperature. Interfacial frictional forces also cause the temperature to rise. Contact with the cold - and often water cooled - rolls causes heat losses. Metallurgical transformations also contribute to temperature changes. A complete mathematical model should account for both thermal and mechanical events which occur in the deformation zone during the rolling process.

Several assumptions regarding material behaviour must be made to allow the formulation of a model. The material is usually assumed to be, and to remain, isotropic and homogeneous; it is considered to be elastic-plastic even though as gross plastic straining takes place, elastic deformations may be quite small in comparison to plastic strains. During forming the volume of the plastic region is taken to remain constant and finally, a plane state of strain is assumed to exist. The equations of motion then

$$\nabla \sigma + \rho \mathbf{b} = \rho \frac{d\mathbf{v}}{dt} \tag{1.1}$$

govern the conditions of equilibrium of both the plastically deforming strip and the elastically loaded work rolls. Here σ is the stress tensor of Cauchy, $\mathbf{b}$ is the vector of body forces, $\mathbf{v}$ is the velocity vector and ρ designates the density of the material.

Plastic flow is according to either the maximum distortion energy criterion of Huber-Mises

$$\frac{1}{2}\sigma_{ij}\sigma_{ij} = k^2 \tag{1.2}$$

or that of Tresca

$$\sigma_I - \sigma_{III} = \sigma_p \tag{1.3}$$

and the appropriate flow rule

$$\dot{\epsilon}_{ij} = \dot{\lambda} \frac{\partial f}{\partial \sigma_{ij}} \tag{1.4}$$

where σ_{ij} stand for the components of the deviator stress tensor and the subscripts I and III denote the maximum and minimum principal stresses, respectively. In equation (1.4) $\dot{\epsilon}_{ij}$ represents the components of the strain rate tensor and $\dot{\lambda}$ is a nonnegative factor of proportionality, given by

$$\dot{\lambda} = \frac{1}{k}\sqrt{\frac{1}{2}\dot{\epsilon}_{ij}\dot{\epsilon}_{ij}} \tag{1.5}$$

and k is the yield strength of the rolled material in pure shear. In equation (1.3) σ_p is the yield strength in tension or compression and is related to k by

$$\sigma_p = \sqrt{3}k \tag{1.6}$$

The plastic potential is designated by f in equation (1.4). In the elastic regions Hooke's law applies.

Appropriately formulated for flat rolling, the equations of motion, yield criterion and the constitutive relations contain five dependent variables, the three stress components σ_x, σ_y and σ_{xy} and the two velocity components v_x and v_y. If ideally plastic behaviour is assumed, that is, $k = \mathrm{constant}$, these equations form the basis of a mathematical model for flat rolling. If material properties are taken to be affected by variables such as strain, strain-rate, temperature, chemical composition and past thermal and deformation history, a choice for an equation relating the flow strength to those variables must be made; testing to determine the flow curves must be conducted and regression analysis is to be used to compute the required parameters.

A solution of the above equations requires the definition of the frictional shear stresses at the roll-workpiece interface. One possibility is to assume that Coulomb friction is present and that the interfacial shear stress equals the roll pressure multiplied by a constant coefficient of friction

$$\tau = \mu s \tag{1.7}$$

Alternatively, the friction stress may be taken as

$$\tau = mk \tag{1.8}$$

where μ is a coefficient of friction and m is a constant friction factor.

Alexander (1972) remarks that in hot rolling, the assumption of sticking friction is often very good; this mathematically would imply that

$$\tau = k \tag{1.9}$$

The roll also deforms during rolling and the shape of the roll-metal interface is not known *a priori*. For the present discussion it is enough to take the function

$$y = f(x)$$

which describes the flattened roll cross-section and which must either be assumed or calculated.

The various contributions to the overall heat balance for a particular cross section in a billet, slab or strip can be summarized as:

(a) heat loss to the environment by radiation and convection along the free surfaces,
(b) heat gain arising from the work of friction forces on the contact surface,
(c) heat gain arising from the work of plastic deformation,
(d) heat loss by the transfer to the roll,
(e) temperature change due to the energy accumulated or realized during the metallurgical transformations.

The heat conduction within the material is described by the general heat diffusion equation

$$\nabla^T (k \nabla T) + Q = c_p \rho \frac{\partial T}{\partial t} \tag{1.10}$$

where k is the heat conduction coefficient, T is the temperature, Q is the rate of heat generation, c_p stands for the specific heat, ρ is the material's density and t designates time.

The rate of heat generation takes into account the plastic work done, energy accumulated in the material due to the increase of dislocation density and heat produced during the softening process. It is described by the equation proposed by Rebelo and Kobayashi (1980)

$$Q = \int_V \left[\sigma_i \dot{\epsilon}_i - \nu A \dot{\epsilon}_i + \nu B \gamma \exp\left(-\frac{D}{kT} \right) \right] dV \qquad (1.11)$$

where V is the volume of the body, σ_i and $\dot{\epsilon}_i$ are the effective stress and effective strain rate respectively, ν gives the energy per unit length of dislocation, A, B are material constants, D is the activation energy of diffusion, γ is the dislocation density and k is the Boltzman constant.

The non-steady state temperature model, with a time-dependent term in equation (1.10), should be applied to all cases of hot rolling of ingots and billets in the roughing mills especially when cooling of the side surfaces and the ends may be significant. During continuous rolling of strips however, a steady state temperature field may be assumed in the roll gap and an additional term connected with heat convection due to the motion of the strip needs to be introduced. A typical steady state convective-diffusion equation is then written as

$$\nabla^T (k \nabla T) - c_p \rho \mathbf{v} \nabla T + Q = 0 \qquad (1.12)$$

In equation (1.12) $\mathbf{v}$ represents the vector of velocities, while all remaining symbols are the same as in equation (1.10).

In order to obtain a solution for the stress, strain, rate of strain, velocity and temperature fields within the rolled strip, appropriate boundary and initial conditions have to be introduced. Both steady-state and nonsteady-state solutions have to satisfy the thermal boundary conditions, which may be of the following types:

(a) the temperature is prescribed along the boundary surface,
(b) the heat flux normal to the surface is prescribed along the boundary,
(c) the linear combination of the temperature and its derivative normal to the surface is prescribed along the boundary.

Schematic illustration of the thermal events including the boundary conditions in the deformation zone is presented in Fig.1.2. Main sources of heat losses include air cooling of the free surfaces and heat transfer to the cooler roll. In modelling, the axis of symmetry and the exit plane from the control volume are treated as insulated surfaces. Finally, constant temperature is prescribed along the vertical plane at the entry to the control volume. The temperature across the contact surface is allowed to be discontinuous, to account for the discontinuity created by the relative velocity between the two surfaces, and also that caused by the resistance to heat transfer across an interface. This latter effect can be specified as a function of the normal stress between the two surfaces in contact or by means of a constant heat transfer coefficient.

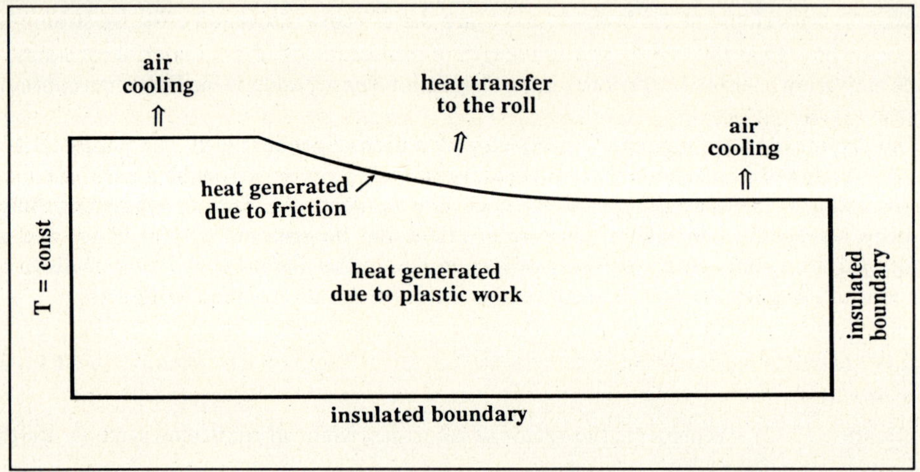

Fig.1.2 Schematic illustration of thermal events and boundary conditions in the roll gap.

Interfacial frictional forces at the roll-strip interface and the shape of that interface need to be accounted for as well. Friction may be expressed in terms of a constant coefficient, multiplying the normal forces or, in terms of another constant operating on the yield strength of the material. In some cases sticking friction may also be assumed to exist. Alternatively, use of experimental data giving the interfacial shear stress as a function of process and material parameters, provides another way to describe the conditions at the contact.

The shape of the contact surface has been determined by researchers employing various techniques. Hitchcock's formula, influence functions, complex variables or stress functions have all been used with reasonable success. Details, including advantages and disadvantages of these schemes will be discussed in Chapter 2.

CHAPTER 2

MATERIAL PROPERTIES AND INTERFACIAL FRICTION

The equations describing the thermal and mechanical events in the roll gap need to be supplemented with appropriate boundary and initial conditions. For the mechanical component these include the resistance to deformation of the metal to be rolled, the frictional forces at the roll-strip interface and the shape of the deformed contact surface. Knowledge of the interface heat transfer coefficient is essential in order to formulate the boundary conditions applicable to the thermal component of the model.

In the present Chapter, two of the above parameters, common to both one-dimensional and two-dimensional models, are described. These are the mechanical properties of the material being rolled and the frictional conditions at the contact surface. The shape of the flattened roll contour is the topic in Section 3.2 while the heat transfer coefficient is discussed in Section 4.4.

2.1 Resistance to Deformation

Some of the difficulties in attempting to predict parameters in hot strip rolling are due to inaccurate determination or inadequate mathematical description of the material's resistance against plastic forming.

The metal's resistance to deformation at high temperatures is affected by the interaction of several metallurgical mechanisms, involving hardening and softening effects which may be examined by concentrating on the history of loading. As straining begins, the grains are progressively flattened and the stresses must be increased to continue the process. At a particular value of strain, dynamic recovery and/or recrystallization may begin. When the rate of softening equals that of hardening a peak in the true stress - true strain curve is observed which is commonly identified by a pair of variables, the peak stress and strain. For many carbon and alloy steels dynamic recrystallization then causes further softening. As straining proceeds, the rate of hardening due to straining and dynamic precipitation may equal and/or overtake the loss of strength. The first event results in steady state behaviour accompanied by grain refinement, while the latter often causes cyclic hardening/softening, resulting in grain coarsening. A thorough review of the mechanics and metallurgy of hot forming has been presented by Sellars (1980). Hardening and softening during deformation have been discussed by McQueen and Jonas (1970) and more recently by Jonas and Sakai (1984).

Techniques to establish high temperature, high rate-of-strain behaviour of metals indicate that the majority of researchers use a cam plastometer to conduct constant strain rate compression tests. Alternatives, however, exist and these as well as the compression tests will be described below (Lenard, 1985).

The tension test is the simplest to perform. The disadvantages associated with it, however, limit its applicability. These include the magnitude of the attainable uniform strain which is significantly less than in other testing methods. As well, when necking develops, the initially uniaxial stress distribution becomes triaxial. Hence, interpretation of the results of the test becomes difficult. A modified tension test was developed by Thomsen et al. (1977) for the purpose of overcoming the cited difficulties. The testing was interrupted at several intervals and the specimens were remachined to remove the necked down sections and to remove the triaxiality. Marion (1978) developed a testing facility for high temperature testing of metals. His paper describes the measurement and control of temperature, temperature gradient and strain. No mention of triaxiality of stress states is made. A split Hopkinson bar was used for high rate tensile testing of aluminium by Nicholas (1981). Strain rates of $10^3 s^{-1}$ were reached. All tests were conducted at room temperature. One of the materials tested was 6061-T651 aluminium alloy which in previous compression tests showed no rate sensitivity at all. In Nicholas' tensile tests the material exhibited measurable rate sensitivity. One of the criticisms often mentioned in connection with the use of split Hopkinson bars in developing data on high rate deformation behaviour is the widely varying rate of straining during the test. A recently introduced pulse shaper by Ellwood et al. (1982) removes this difficulty. As their results show, essentially constant rates of strain are achieved during most of the experiments.

Several researchers use the hot torsion test to determine the flow strength data. As McQueen and Jonas (1970) mention, strains of the order of 20 can be reached. One of the difficulties inherent in the torsion test is the radial variation of the strain rates.

Strains significantly higher than in tension can be achieved in compression testing of solid or hollow cylindrical, conical or flat specimens. Constant strain rate compression tests of solid cylindrical specimens are often performed using a cam plastometer, the operation of which is described in detail by Stewart (1974).

Uniaxial compression testing was used by Suzuki et al. (1968). Their report is very comprehensive, comprising over 100 pages of data for both ferrous and nonferrous metals. They used a cam plastometer and also a drop hammer, measuring the flow stress of 21 nonferrous metals and alloys and 49 steels (carbon, low alloy and stainless) over a wide range of temperatures for each material and for strain rates from $0.1\ s^{-1}$ up to $650\ s^{-1}$.

Modifications to produce constant strain rates on an Instron testing machine, using an analog function generator, were described by Luton et al. (1974). Drop tests, resulting in variable strain rates during deformation, were conducted by Douglas and Altan (1975). Results, obtained by a plane-strain compression test which was developed by Watts and Ford (1952), were discussed by Weinstein and Matsufuji (1968). They concluded that the mean deformation resistance of steel, obtained in plane-strain, variable strain rate tests is different from that obtained from uniaxial, constant strain rate compression. Further, better correspondence of predicted and measured rolling results was obtained by the use of plane-strain material strength data than by the use of uniaxial compression data.

One of the difficulties encountered during compression tests is caused by the friction present between the specimen and the compression platens. There are several ways of accounting for the friction losses. Among these is the one developed by Cook and Larke (1945) who noticed that the contribution of frictional effects is less on slender cylinders than on squat ones. By obtaining results for various height-to-diameter ratios and extrapolating, the zero friction flow strength can be estimated. Alternatively, the analysis developed by Gelin et al. (1981) may be used to calculate the true flow strength.

2.1.1 Empirical Constitutive Relations For Carbon Steels (Karagiozis and Lenard, 1987)

Possessing accurate, reliable and repeatable flow strength data, by itself, is not sufficient if calculations of forces, torques, pressures and powers during bulk or sheet metal forming processes are contemplated. The material's behaviour when subjected to loading must also be represented by an appropriate constitutive equation, giving the strength as a function of other process parameters, such as strain, rate of strain and temperature. This involves selecting the form of the equation as well as using careful nonlinear regression analysis in determining the parameters of that relation. Numerous empirical constitutive equations have been presented in the technical literature. Those specially designed for high temperature, high strain rate applications include the work of Altan and Boulger (1973), Shida (1974), Gittins et al. (1974) and Hajduk et al. (1972). The equation of Ekelund presented by Wusatowski (1969) is also considered, partly for historical reasons.

Altan and Boulger (1973) have compiled a considerable amount of data on the behaviour of engineering materials. For high temperature flow they used the well-known power law

$$\sigma = C\dot{\epsilon}^m \tag{2.1}$$

and presented the results in terms of the strength and strain rate hardening coefficients, designated by C and m, respectively, for each of the metals, for specific temperatures and strains. The chemical compositions of the materials were also given.

One of the most comprehensive set of empirical equations for high temperature behaviour of carbon steels was developed by Shida (1974). The results were obtained using a constant strain rate cam plastometer. The flow strength was given in terms of strain, rate of strain, temperature and carbon content. Shida also compared the number of experiments he conducted to that of others and showed that in each case he performed at least ten times more tests. Shida (1974) expressed the flow strength in the form

$$\sigma = \sigma_f f \left(\frac{\dot{\epsilon}}{10}\right)^m \tag{2.2}$$

where for

$$\overline{T} \geq 0.95 \frac{C + 0.41}{C + 0.32}$$

$$\sigma_f = 0.28 \exp\left(\frac{5}{\overline{T}} - \frac{0.01}{C + 0.05}\right)$$

and

$$m = (-0.019C + 0.126)\overline{T} + (0.075C - 0.05)$$

For

$$\overline{T} < 0.95\frac{C+0.41}{C+0.32}$$

$$\sigma_f = 0.28q(C,\overline{T})\exp\left[\frac{C+0.32}{0.19(C+0.41)} - \frac{0.01}{C+0.05}\right]$$

with

$$q(C,\overline{T}) = 30(C+0.9)\left(\overline{T} - 0.95\frac{C+0.49}{C+0.42}\right)^2 + \frac{C+0.06}{C+0.09}$$

and

$$m = (0.081C - 0.154)\overline{T} - 0.019C + 0.207 + \frac{0.027}{C+0.32}$$

The other parameters are defined as

$$f = 1.3(5\epsilon)^n - 1.5\epsilon$$

and

$$n = 0.41 - 0.07C$$

In the above formulae $\overline{T} = (T+273)/1000$, T is the temperature in ^{0}C, C is the carbon content in weight per cent, ϵ is the strain and $\dot{\epsilon}$ is the strain rate. The author points out that the form of the equation has no physical significance. The range of the applicability of the formulae was given by Shida (1974) as

carbon content	< 1.2%
temperature	700 - 1200 ^{0}C
strain rate	0.1 - 100 s^{-1}
strain	< 70%

Gittins et al. (1974) derived the coefficients of an equation giving the mean yield strength as

$$\sigma = A_0 + \epsilon^{0.2}\left(A_1 + A_2 \ln\dot{\epsilon} + A_3 \frac{T}{1000}\right) \tag{2.3}$$

In this equation the symbol ϵ represents the engineering strain, $\dot{\epsilon}$ is the mean strain rate and T is the temperature.

An equation originally proposed by Ekelund and presented by Wusatowski (1969) gives the yield strength as a function of the temperature, carbon, manganese and chromium contents as

$$\sigma = 9.81(14 - 0.01T)(1.4 + \mathrm{C} + \mathrm{Mn} + 0.3\mathrm{Cr}) \tag{2.4}$$

where C, Mn and Cr are the carbon, manganese and chromium contents, respectively, in weight per cent. While the equation indicates ideally plastic, rate-independent behaviour, Wusatowski (1969) includes it in an empirical relationship which attempts to predict roll separating forces in hot flat rolling. The formula presented by Wusatowski (1969) contains corrections for the rate effects but those corrections do not appear to contribute to the yield strength.

The equation of Hajduk et al. (1972) is given in the form

$$\sigma = \sigma_0 K_T K_\epsilon K_u \tag{2.5}$$

where σ_0, K_T, K_ϵ and K_u are the functions of the material, temperature, strain and strain rate, respectively. Hajduk et al. (1972) give those values for 25 different low and medium carbon steels. A formula identical to (2.5) was also derived by Zyuzin et al. (1964) who presented the coefficients σ_0, K_T, K_ϵ and K_u for a large number of different steels.

The predictive capabilities of these equations are considered next and the results are presented in Figs 2.1 - 2.7. In each, the dependence of the flow strength on strain rate is shown. Two materials, a low carbon and a medium carbon steel are used in the study. The chemical compositions, used in the respective computations, are chosen to be as similar to one another as possible and are indicated below.

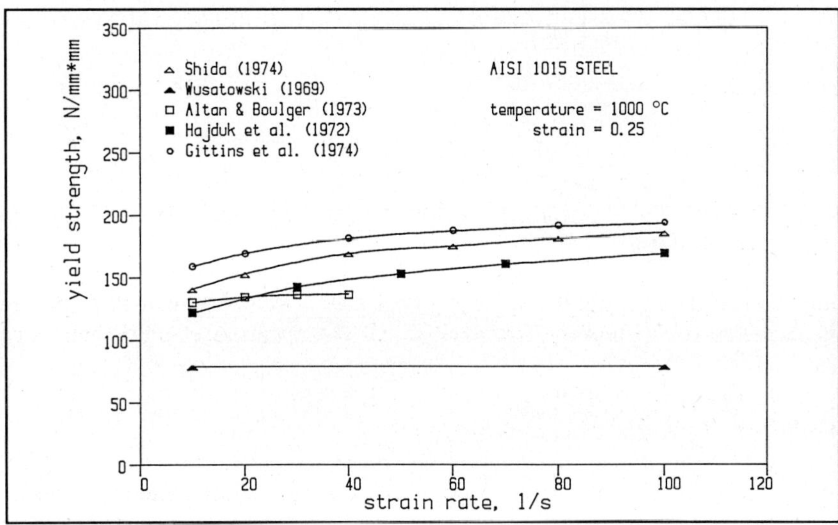

Fig.2.1 Flow curves of AISI 1015 steel at 1000°C and at a true strain of 0.25.

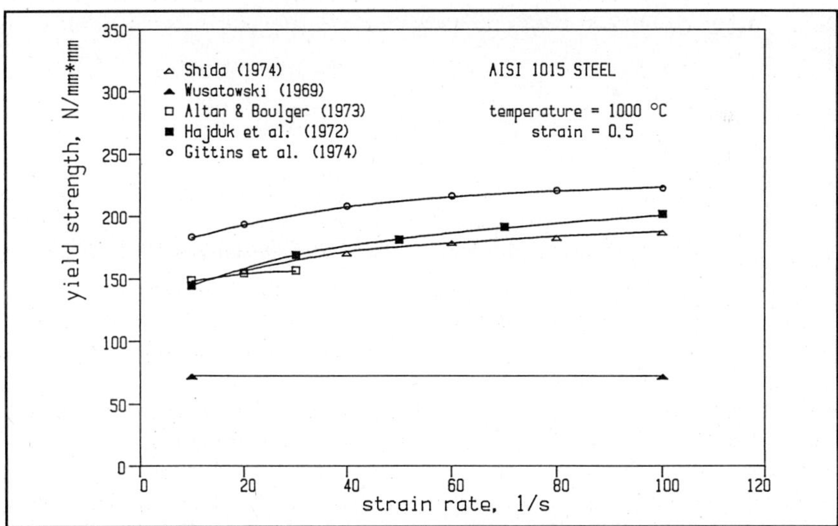

Fig.2.2 Flow curves of AISI 1015 steel at 1000°C and at a true strain of 0.5.

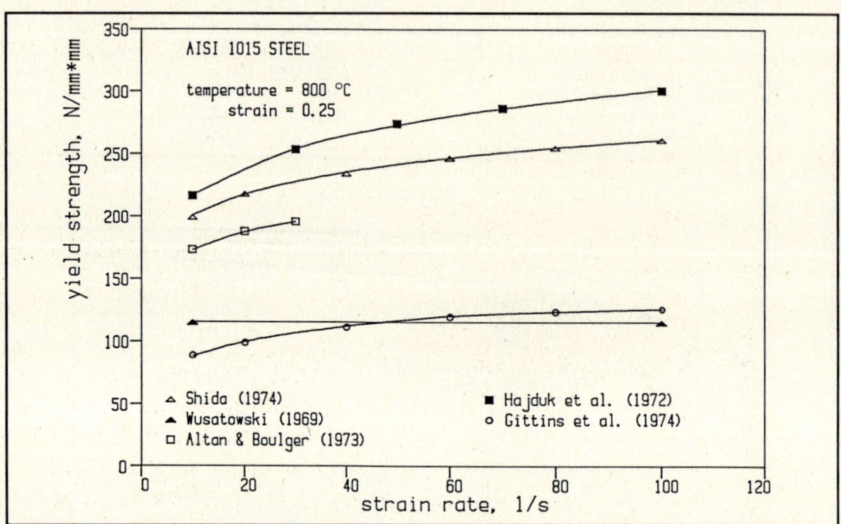

Fig.2.3 Flow curves of AISI 1015 steel at 800°C and at a true strain of 0.25.

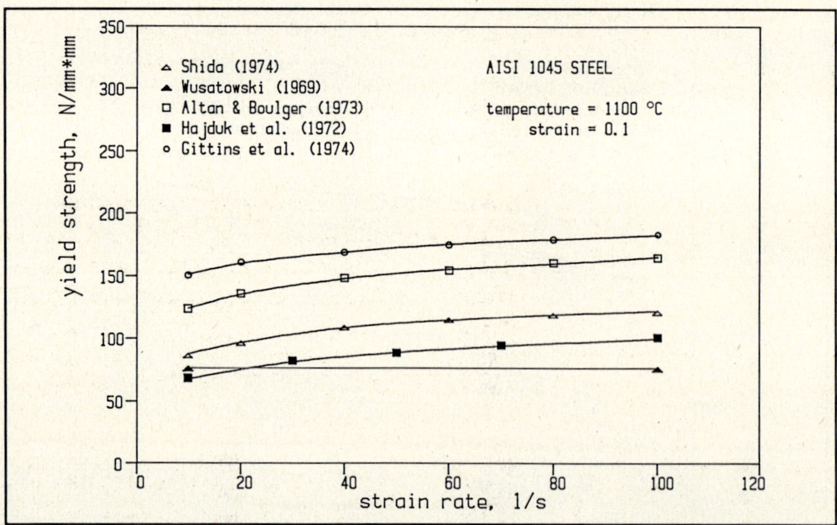

Fig.2.4 Flow curves of AISI 1045 steel at 1100°C and at a true strain of 0.1.

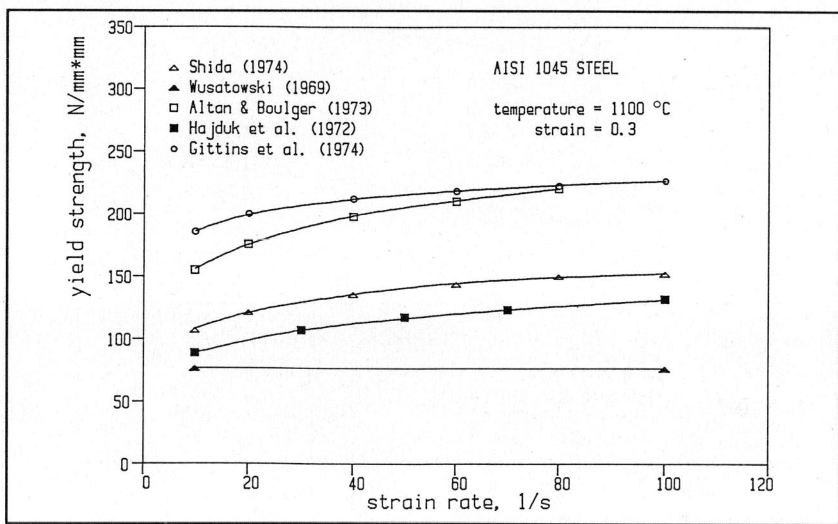

Fig.2.5 Flow curves of AISI 1045 steel at 1100°C and at a true strain of 0.3.

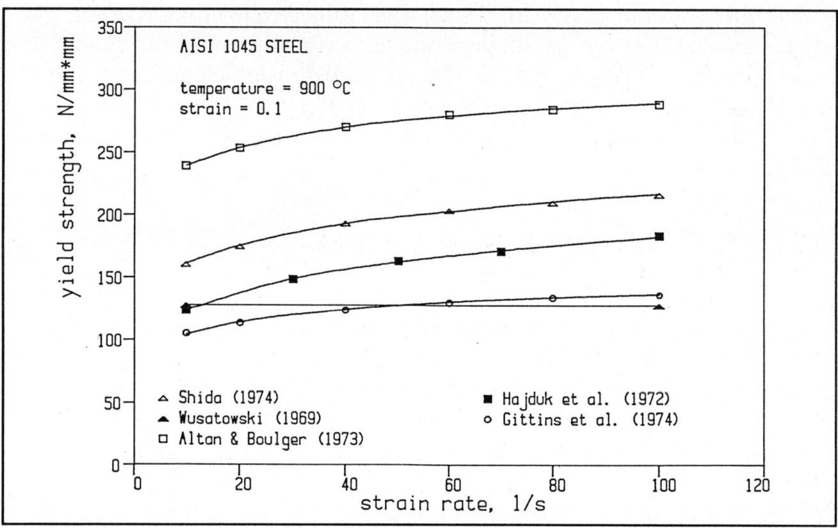

Fig.2.6 Flow curves of AISI 1045 steel at 900°C and at a true strain of 0.1.

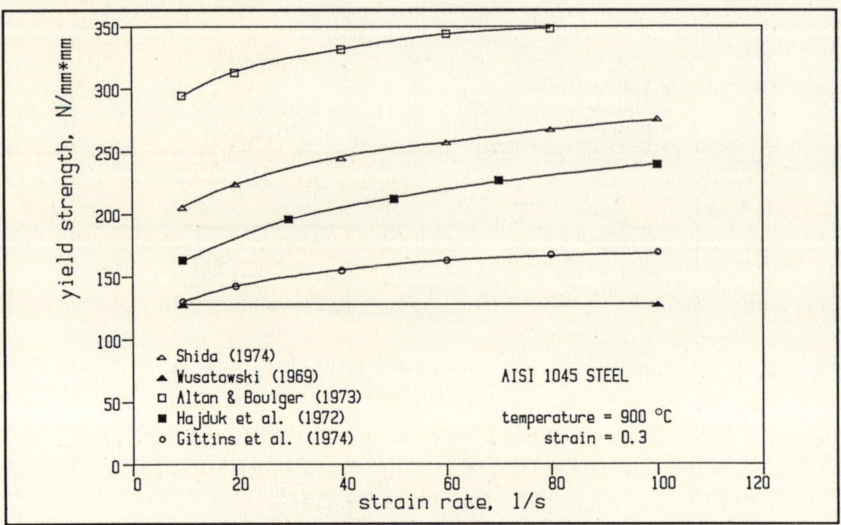

Fig.2.7 Flow curves of AISI 1045 steel at 900°C and at a true strain of 0.3.

The predicted behaviour of AISI 1015 steel when subjected to uniaxial stressing is given in Figs 2.1, 2.2 and 2.3. For the calculations the carbon content is taken as 0.15% and the manganese content as 0.45% by weight. The appropriate coefficients in equation (2.1) are given in Table 2.1. The equation, given by Gittins et al. (1974) for mild steel, which has the chemical composition closest to AISI 1015 with a carbon content of 0.28% and a manganese content of 1.0%, is

$$\sigma = 0.49 + [1 - \exp(-\epsilon)^{0.2}]\left(516.56 + 21.3\ln\dot{\epsilon} - \frac{357230}{T}\right) \tag{2.6}$$

That of Hajduk et al. (1972) for the carbon steel containing 0.13%C and 0.4%Mn, is

$$\sigma = 98.1[17.8\exp(-0.0029T)](1.79\epsilon^{0.252})(0.72\dot{\epsilon}^{0.143}) \tag{2.7}$$

	C, N/mm^2	m
$\epsilon = 0.25$, 1000 °C	117.1	0.045
$\epsilon = 0.50$, 1000 °C	129.5	0.058
$\epsilon = 0.25$, 800 °C	137.1	0.105

Table 2.1 Coefficients in equation (2.1).

It is to be noted that this steel also contains 0.98% Cr and 3.02 % Ni. The equation of Hajduk et al. (1972) for a steel, resembling AISI 1045 and containing 0.43% C, 0.74% Mn, 1.1% Cr and 0.17% Ni, is

$$\sigma = 90.25[20.33\exp(-0.003017)][(1.785\epsilon^{0.248})(0.677\dot{\epsilon}^{0.17}) \qquad (2.8)$$

In Fig.2.1 the flow curves are plotted for a true strain of 0.25; in Fig.2.2 a true strain of 0.5 is used. It is observed that at a temperature of 1000°C the predictions of the five empirical formulae differ broadly. The lowest value of the yield strength are those given by Ekelund whose formula predicts ideally plastic behaviour. The highest strength at both $\epsilon = 0.25$ and at 0.5 is given by Gittins et al. (1974). The differences in predictions grow as the temperature is reduced to 800°C (Fig.2.3). Even at low rates of strain, differences of 200% to 300% are found. Naturally, differences are to be expected since the chemical compositions of the steels for which the above formulae were derived are not identical. It is reasonable to expect the predictions of Gittins et al. (1974) to be the highest since their mild steel contains almost twice the carbon and manganese as the others. Also, since Ekelund's relation contains only temperature as the independent variable, its low flow strength value is not surprising. It is unexpected, however, to observe that Gittins et al. (1974) give lower strength for lower temperatures, which is of course contradictory to observations. It is the negative value of their A_3 coefficient, divided by the temperature, which is responsible. Further, while in the investigations of Altan and Boulger (1973), Shida (1974) and Gittins et al. (1974) the chemical compositions of the metals are reasonably close to one another, differences of 20% to 25% magnitude at T = 1000°C and of 35% at T = 800°C are observed. The temperatures were chosen arbitrarily for the computations presented above. It is, of course, recognized that the Ar_3 transformation point for an AISI 1015 steel is 820°C - see Engineering Properties of Steel (1982) - and precisely because of this the comparisons shown in Fig.2.3 were done at 800°C.

The flow curves for AISI 1045 steel are given in Figs 2.4 and 2.5 for 1100°C and in Figs 2.6 and 2.7 for 900°C. Again, the information presented does not allow one to choose any of the published empirical formulae for further calculations with any confidence.

2.1.2 High Strength Low Alloy Steels (Lenard and Karagiozis, 1989)

Consideration of published data on the mechanical response of HSLA steels also reveals some contradictions. These may be observed in Table 2.2 where a compilation of available results on the strength of some niobium bearing steels is presented. Peak strain, corresponding peak strength, strain rate, temperature and chemical compositions are given along with the mode of testing and prior solution heat treatment. As is evident, testing techniques, heat treatment, process parameters and results vary broadly.

Stewart (1975) and Armitage et al. (1976) use a cam plastometer to produce constant strain rate compression. Sankar et al. (1979) use a microprocessor controlled hot torsion testing device as do Gittins et al. (1974). Tension tests are conducted by Wilcox and Honeycombe (1984) and by Maki et al. (1981). Compression of axially symmetrical specimens is carried out by Tiitto et al. (1983) and by Bacroix et al. (1981). The results of D'Orazio et al. (1985) are also given in Table 2.2. Parameters used by various researchers differ widely, making direct comparison of results difficult. Inconsistency of the values is still noticeable. For example, for 0.038% Nb bearing steel, Tiitto et al. (1983) and Maki et al. (1981) give data that differ by approximately 100%. As well, peak strain values appear to depend on the testing method in a significant manner. The only area of agreement concerns the retardation of recrystallization caused by the addition of niobium.

2.1.3 Representation of the Results of Compression Tests (Lenard et al., 1987)

Compression tests have been widely used to evaluate the material's yield strength, which is used further in various mathematical models of the rolling process. Some typical examples of application of compression tests to evaluate yield strength of low carbon steel and vanadium steel are presented in this section. The results of the experiment will be used in the following Chapter in calculating the roll force and the mill power. Three techniques to model material behaviour are employed:

(a) development of a multidimensional databank,
(b) use of average values,
(c) development of empirical equations.

Nb	C	Mn	T °C	$\dot{\epsilon}$ s^{-1}	σ_p N/mm^2	ϵ_p	test	solution treatment °C/min	Ref.
0.08	0.16	1.26	900	0.5	154	n/a	compression	1200/60	Stewart (1975)
			997		88				
0.06	0.10	1.65	949	15.0	154	0.25	compression	1290	Armitage et al. (1976)
0.05	0.12	0.94	900	0.1	144	0.75	torsion	1250/30	Sankar et al. (1979)
			950	0.1	121	0.65			
			1000	0.1	104	0.56			
			900	1.0	185	0.80			
			1000	1.0	125	0.68			
0.046	0.05	0.35	1000	7.0	160	n/a	torsion	1100/10 1300/10	Gittins et al. (1977)
0.04	0.25	1.15	850	0.0014	79	n/a	tension	1300/30	Wilcox & Honeycombe (1984)
			900		62				
			950		42				
			1000		35				
0.035	0.05	1.25	900	0.00037	73	0.44	compression	n/a	Sakai et al. (1981)
			1075	0.037	63	0.26			
0.038	0.08	1.25	900	0.006	117	1.2	compression	1200/30	Tiitto et al. (1983)
			1000	0.006	75	0.3			
0.038	0.12	1.4	900	0.005	70	0.33	tension	1100/1	Maki et al. (1981)
			1000		35	0.18			
			1100		28	0.15			
0.035	0.05	1.25	925	0.014	100	n/a	compression	n/a	Bacroix et al. (1981)
				0.0014	75				
0.02	0.09	0.90	750	0.2	152	0.54	compression	1150/2	D'Orazio et al. (1985)
				1.0	190	0.62			
				10.0	206	0.64			
			850	0.2	110	0.52			
				1.0	125	0.60			
				10.0	150	0.63			
			950	0.2	65	0.50			
				1.0	80	0.56			
				10.0	100	0.62			
0.01	0.16	0.56	1000	7.0	160	n/a	torsion	1100/10 1300/10	Gittins et al. (1977)

Table 2.2 Compilation of published constitutive results for niobium steels.

The materials used in the study are low carbon and vanadium steels obtained, in the form of hot rolled sheets, from Sidbec-Dosco Ltd. of Canada. The chemical composition of the steels in weight per cent is given in Table 2.3. The nitrogen content for both steels is 80 ppm.

steel	C	Mn	Al	S	P	Cu	Ni	Cr	Si	Mo	V
#1	0.18	0.8	0.05	0.015	0.015	0.1	0.05	0.05	0.03	0.005	-
#2	0.15	0.65	0.04	0.015	0.015	0.1	0.05	0.05	0.03	0.005	0.035

Table 2.3 Chemical composition of the materials used in the experiment.

The experiments were carried out on a microprocessor-controlled servohydraulic testing system. By providing an exponentially decaying voltage to the servovalve, the actuator travels to cause constant strain rate throughout its stroke. Strain rates up to 50 s^{-1} are reached, using specimens not higher than 20 mm. The dynamic response of the system is adequate. Even at high speeds, when the time taken to overcome the inertia of the servovalve is significant, command and feedback signals differ no more than 10%. Data acquisition is achieved by the A/D converter of the computer and the resulting true stress - true strain curves are obtained directly on the attached plotter. Further, to simulate the conditions in the finishing train of a hot strip mill, interrupted compression testing up to 10 stages can be conducted, within a highly stable, resistance furnace. Strains, rates of strain and unloading times may be selected per cycle, in an interactive manner, at the keyboard of the computer. Details of the testing facility have been published by Lenard (1985).

Cylindrical samples of 12.5 mm diameter and 19.05 mm height were machined with the longitudinal axis of symmetry in the direction of rolling. All samples were annealed at 1000°C for 2 hours, air cooled and austenitized at 1200°C for 30 minutes, followed by quenching in iced brine. Both treatments were conducted in stainless steel envelopes, filled with zirconium shavings. Delta Glaze 19 glass lubricant was used to reduce friction during the compression test.

Single stage true stress - true strain curves at 900°C and at various strain rates are given in Fig.2.8 for the low carbon steel and in Fig.2.9 for vanadium steel. For lower rates of strain the flow curves of both steels reach a plateau indicating the initiation of dynamic recrystallization followed by softening. At $\dot{\epsilon} = 50\ s^{-1}$ or beyond neither steel has sufficient time to reach the level of straining for dynamic recrystallization to begin. The dependence of the peak strengths and strains for the two steels on the rate of straining may be inferred from Figs 2.8 and 2.9. The ability of vanadium to retard the recrystallization, at least at higher rates, is observed.

In order to simulate the steels' behaviour during several stages of rolling, multistage compression tests were conducted. The flow curves resulting from these are shown in Figs 2.10, 2.11 and 2.12 for the carbon steel and in Figs 2.13, 2.14 and 2.15 for the vanadium steel.

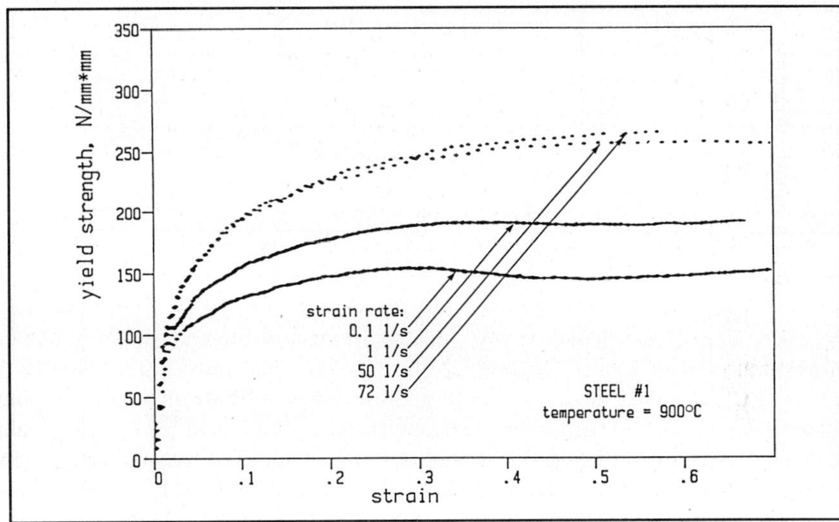

Fig.2.8 Flow curves of the low carbon steel #1.

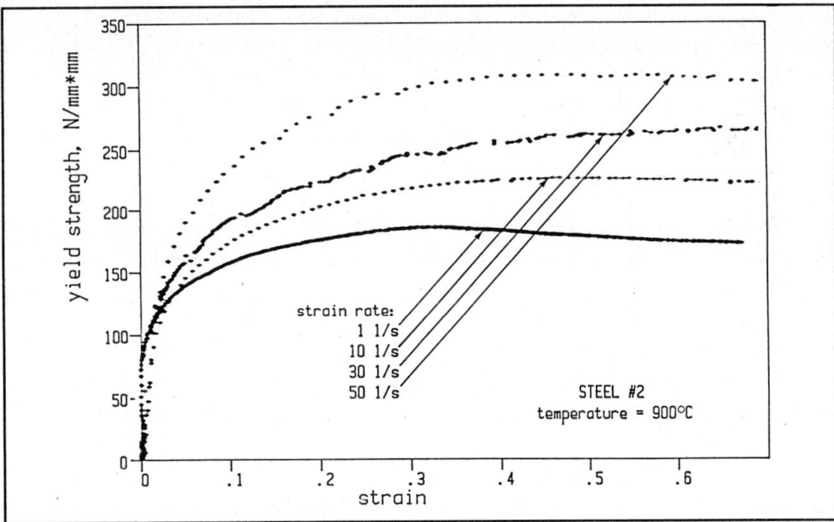

Fig.2.9 Flow curves of the vanadium steel #2.

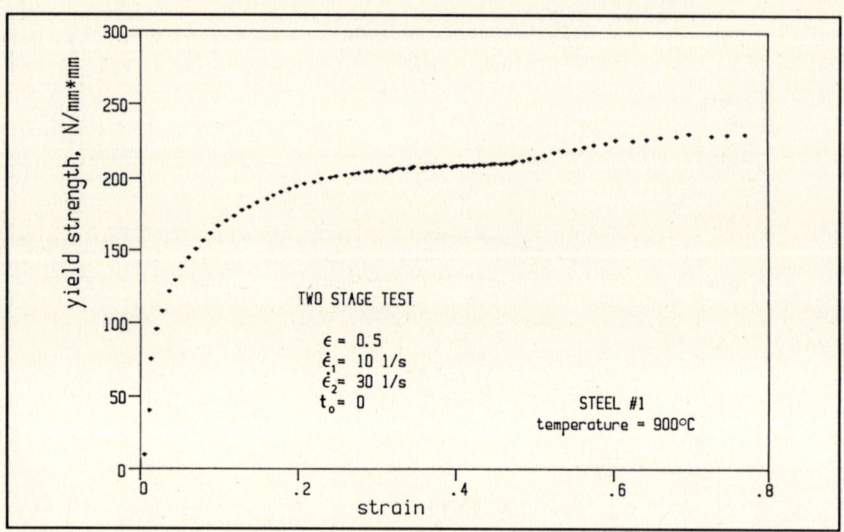

Fig.2.10 Two stage flow curve of the low carbon steel #1.

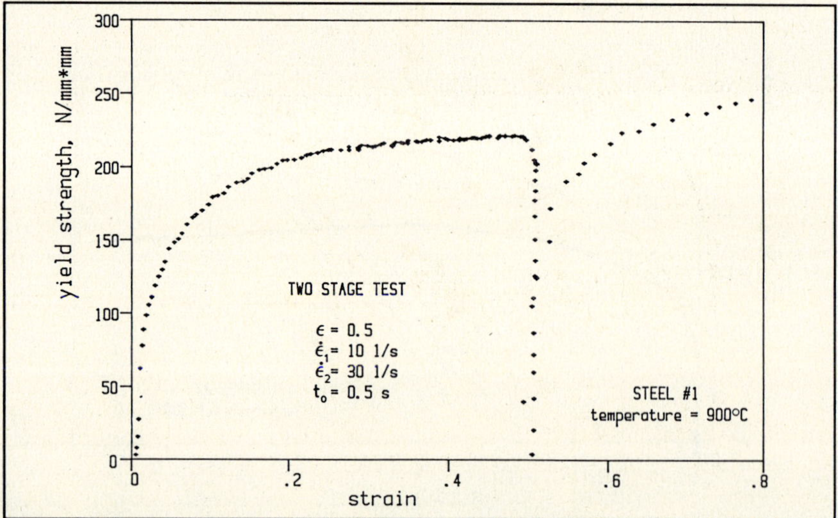

Fig.2.11 Two stage flow curve of the low carbon steel #1.

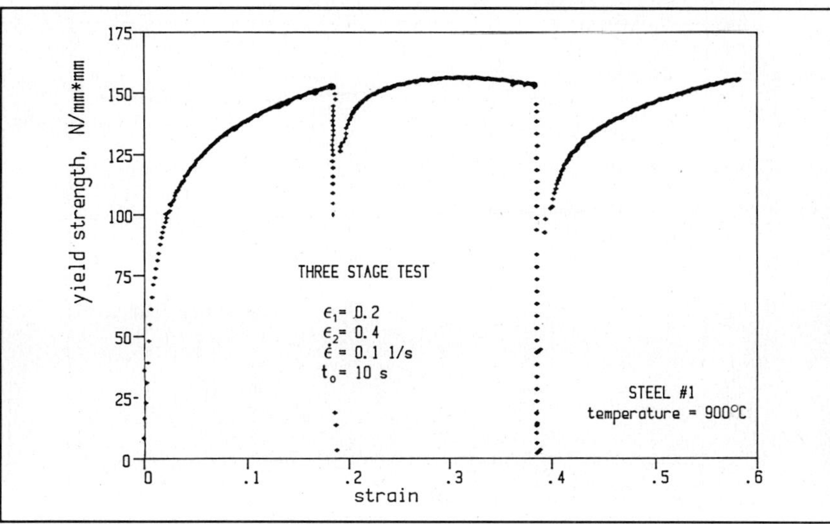

Fig.2.12 Three stage flow curve of the low carbon steel #1.

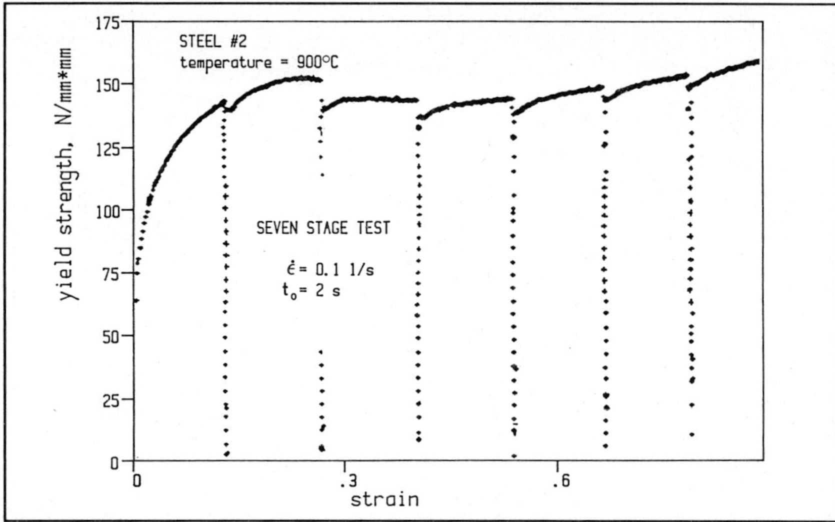

Fig.2.13 Seven stage flow curve of the vanadium steel #2.

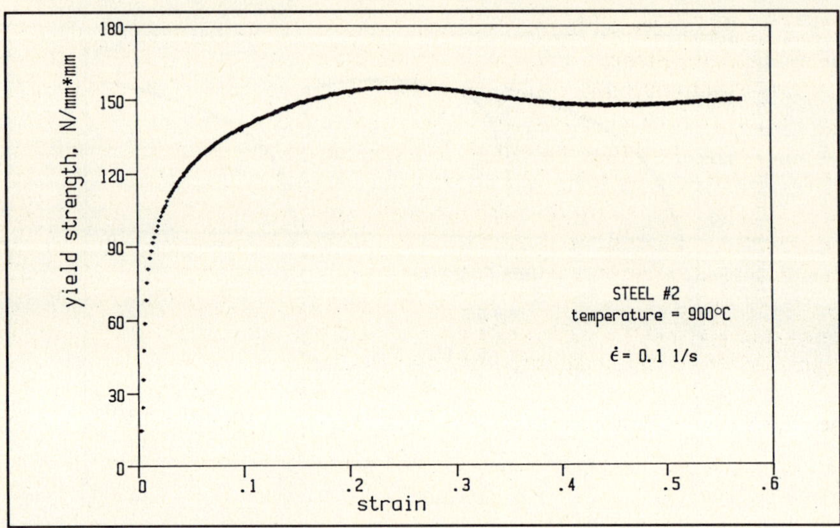

Fig.2.14 Single stage flow curve of the vanadium steel #2.

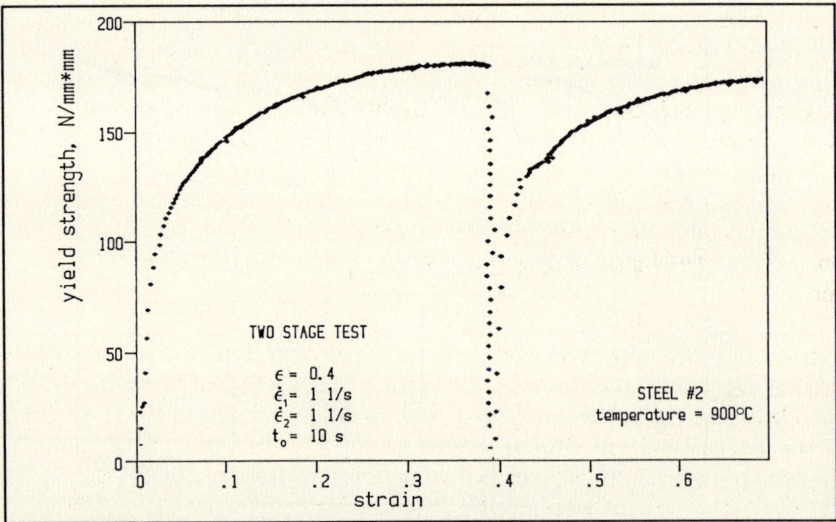

Fig.2.15 Two stage flow curve of the vanadium steel #2.

The strain rate was increased from 10 s^{-1} to 30 s^{-1} at $\epsilon = 0.5$ without unloading and, as expected, the steel's resistance to deformation increased accordingly, as seen in Fig.2.10. When a 0.5 s delay was introduced at the interruption, some fractional softening, but essentially unchanged behaviour was observed (Fig.2.11). The results of a three stage experiment conducted at $\dot{\epsilon} = 0.1\ s^{-1}$ are shown in Fig.2.12. The test was interrupted at strains of 0.2 and 0.4 for 10 seconds in both stops. It is noted that in spite of significant fractional softening at both unloadings, the single stage flow curve for the same strain rate (Fig.2.8) envelopes the one in Fig.2.12. Essentially the same conclusions, within permissible experimental errors, may be drawn when comparing the seven stage flow curve of the vanadium steel, shown in Fig.2.13 and the uninterrupted one, given in Fig.2.14. Further, the two stage test, demonstrated in Fig.2.15 compares well to the single stage curve of Fig.2.9, in spite of the 10 s interruption. In what follows therefore, the single stage results are analyzed.

As mentioned above, flow curves may be represented by an equation of the general form

$$\sigma = f(\epsilon, \dot{\epsilon}, T)$$

where ϵ is the true strain, $\dot{\epsilon}$ is the strain rate and T is the temperature. The choice of the specific form of the equation is a difficult and important one. Three methods are followed in this analysis. First, a multidimensional databank is created, then empirical constitutive relations are developed and finally, the use of the average flow strength is investigated.

The true stress - true strain data, collected during the actual test in digital form is used in the multidimensional databank. A multidimensional matrix, having the strain, rate of strain and temperature as coordinates, is established, giving the values of the material's strength for specific ϵ, $\dot{\epsilon}$ and T values. When the flow strength at points other than those stored but within the existing domain, are required, linear interpolation is employed to compute them. Extrapolation allows determination of the strength when points outside the domain are considered.

It is well known that constitutive behaviour of steels at high temperatures allows one to identify three regions of a stress - strain curve. These include the initial hardening part followed by softening and then either by steady state flow or cyclic hardening/softening. An empirical relation should incorporate all of these phenomena or, at the very least, should reproduce the $\sigma - \epsilon$ curve up to the strain experienced in the forming process. For the rolling passes of concern in this project strain levels reached the softening region only. Hence, it is sufficient to use the equation developed by Halal and Kaftanoglu (1984) of the form

$$\sigma = \sigma_{po}(1 + B\epsilon)^n - A\epsilon^m \tag{2.9}$$

where σ_{po}, B, A, n, m are material constants. Equation (2.9) in this form allows for both strain hardening and softening. Using the single stage constitutive data and nonlinear regression analysis leads to the results given in Table 2.4 for both steels.

The assumption of ideally plastic behaviour in metal forming analysis often reduces the complexity of the problem albeit at the expense of accuracy. In order to test the magnitude of that loss the average flow strength, defined as

$$\bar{\sigma} = \frac{1}{\epsilon}\int_0^{\epsilon} \sigma d\epsilon \tag{2.10}$$

is calculated for each pass.

These results are used in the following chapter together with various one-dimensional models in the calculation of the roll force and power.

2.2 Friction in the Flat Rolling Process (Lim and Lenard, 1984; Karagiozis and Lenard, 1985)

The boundary conditions to be satisfied by the solution of the equations of the mechanical component of the model of the flat rolling process include the friction stress at the roll-strip interface as well as the shape of that interface. The latter will be treated in some detail in Section 3.2. The purpose of the present section is to discuss the former boundary condition, the interfacial friction, which is probably the least well understood and the least researched phenomenon.

2.2.1 Friction as a Function of Process and Material Parameters

The traditional approach in modelling has been to assume that Amonton's law holds and that friction forces in the roll gap are proportional to the normal forces, with the factor of proportionality - the coefficient of friction - depending only on the contacting materials. This implies that an average frictional coefficient should be determined for each particular set of parameters, since, as pointed out by Rabinowicz (1965), this coefficient depends critically on a large number of variables. According to him the adhesion hypothesis, stating that friction is due to adhesive forces between the contacting surfaces, is almost universally accepted by

material	$\dot{\epsilon}$ s^{-1}	σ_{p0} N/mm^2	B	n	A N/mm^2	m
low carbon steel	0.1	80	43.13	0.5403	546.03	0.9086
	1	90	52.25	0.5162	556.20	0.8524
	26	101	54.21	0.5030	560.65	0.9231
	50	101	60.61	0.4981	561.64	0.8616
	72	101	65.23	0.4937	563.21	0.8382
vanadium steel	1	98	48.86	0.4134	502.10	0.8449
	10	99	46.08	0.4125	416.88	0.8893
	30	100	50.88	0.4719	433.20	0.9182
	50	101	63.82	0.4633	448.82	0.8188

Table 2.4 Material constants in the constitutive equation (2.9).

the scientific community. Further, Rabinowicz (1965) lists and discusses the parameters that influence surface interactions. These include the interfacial velocity and volume properties, such as the yield strength, flow strength, penetration hardness, strain hardening sensitivity, Young's modulus, shear modulus, elastic energy, ductility and thermal properties. Further, surface properties, such as chemical reactivity, the tendency to absorb molecules from the environment, surface and interfacial energy must also be considered. The work of Mokhtar et al. (1980) also indicates the dependence of the coefficient of friction on atomic volume, surface energy, hardness, recrystallization temperature, melting point, boiling point, and coefficient of thermal expansion.

Tabor (1981) emphasized that the three major elements of the frictional process are the true area of contact, the nature of strength of the interfacial bonds and the way in which the contacting materials are sheared and ruptured. In his conclusions Tabor (1981) pointed out areas that need further research. Regarding sliding in deformation processes, work-hardening which is a consequence of sliding, needs attention. As well, there appears to be no reliable experimental method of determining the true area of contact. Schey (1980) stressed the need to realize the importance of material properties when the nature of friction between contacting surfaces is studied. Further, in discussing dry friction, he mentioned the significance of the proper choice of die composition and/or surface cooling. Plevy (1979) considered the role of friction in metalworking with the objective of reducing the energy losses. The influence of loading and sliding velocities on the bulk body was investigated by Al-Rubeye (1980). The effects of asperity height, wavelength, and angle of inclination on the coefficient of friction were studied by Koura and Omar (1981). Kuhlmann-Wilsdorf (1981) proposed a hypothesis regarding the observed significant increase of the coefficient of friction with rising interfacial pressure. She suggested that this phenomenon may be the result of the grouping of several contact spots. A thorough review of experimental work on contact of rough surfaces was given by Woo and Thomas (1979). As they state, there appears to be a consensus among researchers regarding the relationship of real area of contact, distribution of contact spot sizes and the separation of surfaces and the load. Mokhtar (1981) concluded that the coefficient of friction is an intrinsic property of metals. The mean frictional coefficient was measured in several metal forming processes: Tabata and Masaki (1978) considered forging, while Kaftanoglu (1973) and Ghosh (1977) considered deep drawing and stretch forming. Theoretical models of friction were presented by Klamecki (1980), Phan-Thien (1981) and Zmitrowicz (1987). Wanheim and Bay (1978) also developed models for interfacial friction forces applicable specifically in the flat rolling process.

The long list of variables emphasizes the suspicion of early workers that the coefficient of friction in the roll gap does not remain constant, a fact that now has been demonstrated to be true in several instances. Analytical work, incorporating a variable frictional coefficient was begun in 1939. Nadai (1939) calculated the effect of various frictional conditions on the roll pressure distribution. He used three hypotheses. First, he assumed that the friction coefficient μ = constant, then he considered the friction force to be constant, and finally he assumed that surface friction is proportional to the relative velocity of the strip and the work roll. Comparing the results of the last assumption to the measurements of Siebel and Lueg (1933), Underwood (1950) concluded that Nadai's last hypothesis is probably valid. Brown's investigation of the variation of the coefficient of friction is described by Underwood [48]. He calculated the distribution reported by Siebel and Lueg (1933) and he found that this requires the coefficient of friction to vary considerably in the roll gap, having a value of about 0.3 at entry and changing to -0.2 at exit.

A torque-force-slip model of cold strip rolling was derived by Hsu (1980). His work is referenced by McPherson (1974); however, details of the actual model were published as recently as 1980. Hsu assumed that the interfacial shear forces depend on the magnitude of relative velocity between the work roll and the strip. He used the influence functions derived by Jortner et al. (1958) to calculate roll deformation.

The experimentally determined distributions of frictional coefficients of Al-Salehi et al. (1973) were incorporated in a model by Lenard (1981). In this study, roll deformation was analyzed by a finite element routine and it was shown that the technique reproduced roll pressure distributions, roll forces and roll torques substantially better than the traditional technique of Orowan (1943).

Experimental verification of the variation of μ in the roll gap can also be found in the technical literature. In one of the earlier projects, van Rooyen and Backofen (1957) used two pin-type transducers imbedded in the work roll of 152.4 mm diameter to measure simultaneously normal pressures and interfacial shear stresses. Aluminium strips were rolled with no lubrication and the ratios of surface shear stresses to normal stresses were plotted for a sand-blasted roll with a roughness of 1.52 μm RMS. As well, the measurements were repeated introducing rolling oil in the roll gap and using the ground roll. Reductions of 51%, 52% and 54% respectively, were obtained. The results showed significant variation of the coefficient of friction when the sand-blasted roll was used or when the ground roll was lubricated. Using the ground roll without lubrication resulted in a distribution of the frictional coefficient that was essentially flat on either side of the neutral point. In all three cases presented by the authors the location of the zero interfacial shear appeared to coincide with the location of the maximum roll pressure. As well, single, smooth pressure peaks were reported. The constrained yield strength of the material was also shown and some strain hardening was observed.

Rabinowicz (1965) pointed out the importance of removing all surface layers from the contacting surfaces by the use of strong caustic soda solutions. In their tests, van Rooyen and Backofen (1957) used a 5% NaOH solution to clean both rolls and strips; further, the roll surfaces were washed and cleaned with carbon tetrachloride.

A cantilever type transducer machined from the roll was employed by Banerji and Rice (1972) and the results obtained during 12% reduction of aluminium strips were reported. Again, the friction coefficient varied widely throughout the roll gap but the location of the single pressure peak did not coincide with the location of the neutral point. Al-Salehi et al. (1973) presented results obtained by rolling aluminium, copper and mild steel strips and using a technique similar to that of van Rooyen and Backofen (1957). Normal pressure, interfacial shearing stress and coefficient of friction plots were presented for 14.4% and 34.4% reduction of aluminium strips, for 14.4% and 17.2% reduction of copper and for 7.3% reduction of mild steel. Two rolls of 79.4 mm radius were used with average surface roughness of 0.2 to 0.37 μm CLA. In all five instances the variation of the coefficient of friction in the roll gap was very pronounced. Two pressure maxima were recorded during 14.4% reduction of aluminium and three pressure maxima were recorded during 34.41% reduction of the same material. Single pressure peaks were evident in the other three experiments. Al-Salehi et al. (1973) concluded that "... rolling theory can account only for the portion of the results ...".

Cole and Sansome (1968) reviewed the techniques used in measuring interfacial normal and tangential forces. As they point out, one of the major difficulties in using any transducer embedded in the work roll, is the possibility of the rolled strip extruding into the clearance between the transducer and the roll. The magnitude of this clearance is, of course, dependent on the magnitude and direction of the deformation of the transducer during the measuring cycle. MacGregor and Palme (1948) used a pin-type transducer of 2.35 mm diameter and a radial clearance of 0.097 mm while van Rooyen and Backofen (1957) used a precision fit and the radial clearance of their transducer was 0.005 mm. MacGregor and Palme (1948) succeeded in measuring the normal pressure only because back extrusion prevented recording of the tangential forces. In the three component dynamometer of Muzalevskii and Grishkov (1959) a radial clearance of 0.041 mm was used but no mention of back extrusion is made. The cantilever-type transducer of Banerji and Rice (1972) had a small clearance whose magnitude is not reported by the authors. However, no significant back extrusion was evident during testing. A revised version of that cantilever was presented by Jeswiet and Rice (1982), while a further revision was developed by Britten and Jeswiet (1986).

A general and inescapable conclusion emerges following a review of the published literature, indicating that the choice of an appropriate value of the coefficient of friction for use in modelling of metal forming processes is still problematic. The data presented below may offer some guidance in making that choice.

2.2.2 Interfacial Forces in the Roll Gap During Flat Rolling

The results of a set of experiments on cold rolling with and without lubrication, and on dry warm rolling will be presented in this section. Several aluminium alloys were used in the tests.

Experimental equipment. The friction experiments were conducted on a laboratory two-high mill built in the machine shop of the Faculty of Engineering at the University of New Brunswick. The 254 diameter, 100 mm wide work rolls were made of FNS tool steel containing 1.5%C, 0.35%Mn, 0.3%Si, 11.5%Cr, 0.75%Mo and 0.8%V that was especially developed by Atlas Steel Co. Ltd for the rolls of cold rolling mills. The rolls were hardened to $R_c = 48$ and ground to a finish of 0.2 μm CLA. The rolls were driven by a rheostat-controlled, variable-speed 25 kW DC motor, through a ten-speed gear box, a pinion box and a pair of universal joints. Torque transducers were placed in the spindles of the universals and two force transducers were located under the bearing blocks of the lower roll to allow measurement of the vertical component of the roll separating force. The bearing blocks were carefully greased in their guides so that frictional forces, resulting from the horizontal component of the separating force could be safely ignored.

Screw-down was controlled by two worm-gear drives, both connected to the bearing blocks of the upper work roll. One turn of the screw-down control resulted in a 0.025 mm change of the roll gap. The mill frame had a total load-carrying cross sectional area of 20.6 x 10^3 mm^2. During the design phase this was expected to cause the mill to be fairly rigid. Mill stretch, however, was measured to be approximately half a millimeter during 15 percent reduction of 1100-T0 aluminium strips.

The roll gap forces were measured by 1.6 mm diameter pins, made of drill rod material, hardened and ground to be flush with the work roll surface. The pins pressed directly on force transducers whose signals were measured through a mercury bath slip ring. The results were collected by a microcomputer.

Two locations were used in placing the pins and transducers in the rolls. In the first, shown in Fig.2.16, the pins followed one another and were located on the centre line of the top work roll. In the second, shown in Fig.2.17, they were located side by side, 25 mm apart. One of the pins was oriented with its axis of symmetry in the radial direction and the other one was tilted 25° from the first. A complete test consisted of two runs, one with the rolls rotating in a certain direction followed by the next, in reverse. Four sets of signals were collected and these, used in a force balance, yielded the interfacial frictional forces, roll pressures and their ratio the frictional coefficients. The details of the derivation are given by Lim and Lenard (1984).

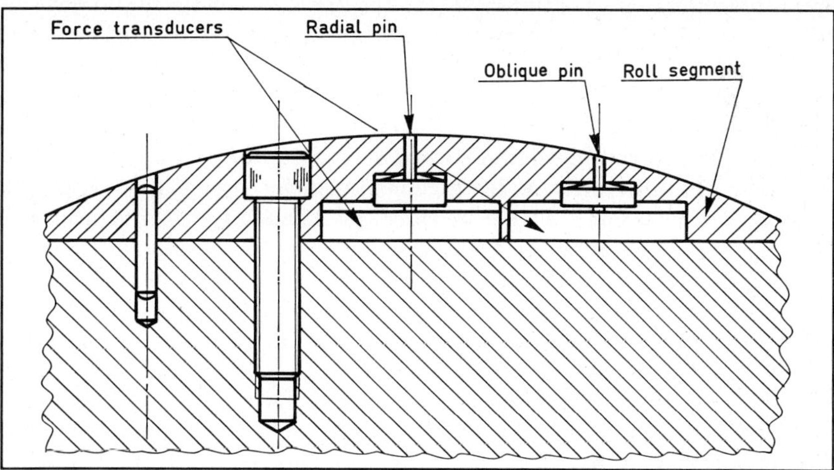

Fig.2.16 Pins located on the centre line of the top work roll.

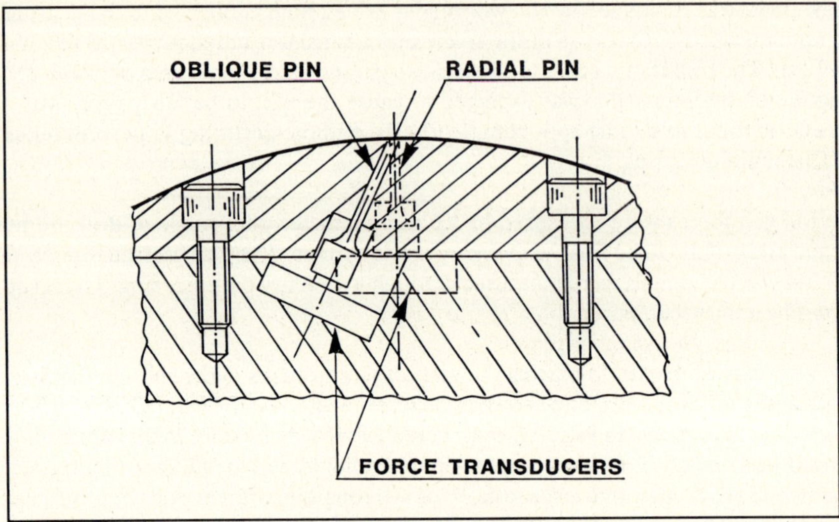

Fig.2.17 Pins located side by side, 25 mm apart.

Materials. Various aluminium alloys were used in the experiments. Their chemical compositions in weight per cent are given in Table 2.5. The flow curves of the metals have also been determined. They, along with other relevant physical properties, are shown in Table 2.6.

Cold rolling without lubrication. Experimentally determined roll pressure and coefficient of friction distributions obtained during cold rolling of 1100-T0 aluminium strips are given in Figs 2.18 - 2.21. In each case the distance along the roll gap, measured from entry, is plotted on the abscissa. Roll pressures and the frictional coefficients are plotted on the ordinates. The percentage reduction and roll rpm are also given in each figure. Roll velocity is kept very low; 0.35 rpm is used for 10.51%, 14.32% and 22.35% reductions. In all four experiments significant variation of the coefficient of friction is evident. In general, the coefficient appears to increase as reduction is increased. In the experiments shown in Figs 2.18 - 2.20, the location of zero friction, i.e. $\mu = 0$, appears to be very close to the location of the maximum pressure.

In these tests the first set of transducers, with the pins following one another, was used. In all of the subsequent experiments the configuration of Fig.2.17 was employed. As well, in the following plots, in addition to the coefficient of friction and normal pressure distributions, the variations of the interfacial shear stress have also been given.

	Mn	Si	Al	Fe	Cu	Mg	Cr	Zn	Ti
Al 1100 T0 & H14	0.05	1.0	rest	-	0.05-0.2	-	-	0.1	-
Al 2024 T3	0.3-0.9	0.5	rest	0.5	3.8-4.9	1.2-1.8	0.1	0.25	0.15
Al5052 H34	0.1	0.25	rest	0.4	0.1	2.2-2.8	0.15-0.35	0.1	-
Al 6061 T6	0.15	0.4-0.8	rest	0.7	0.15-0.4	0.8-1.2	0.04-0.35	2.5	0.15

Table 2.5 Chemical composition of the strip materials, weight %.

	yield strength N/mm^2	ultimate strength N/mm^2	hardness Vickers	r value	constitutive law
Al 1100 T0	-	-	30	0.871	$\sigma = 85(1 + 96.8\epsilon)^{0.2}$
Al 2024 T3	275.6	420.3	143	0.946	$\sigma = 299.9(1 + 86\epsilon)^{0.123}$
Al 5052 H34	179.1	234.3	95	1.446	$\sigma = 199.6(1 + 201.8\epsilon)^{0.097}$
Al 6061 T6	241.2	289.4	128	0.641	$\sigma = 265(1 + 192.1\epsilon)^{0.029}$

Table 2.6 Mechanical properties of the strip materials.

In Fig.2.22, for 8.87% reduction of the Al 2024 - T3 strip, at roll rpm of 4, the frictional coefficient is essentially constant from entry until the neutral point is reached. For Al 1100 - T0 and Al 1100 - H14, rolled at 0.35 and 10 rpm respectively - see Figs 2.19 and 2.23 - the frictional coefficients vary widely from entry to exit. As well, the roll pressure distributions are not smooth and double peaks are apparent in both instances. In Figs 2.19 and 2.23, neutral points and pressure maxima are at the same angular position. These observations along with the results of Lim and Lenard (1984), mentioned above, reinforce the conclusion that the assumption of the existence of a constant frictional coefficient in cold strip rolling is invalid.

Effect of reduction and entry velocity. The effect of per cent reduction on the average coefficient of friction is shown in Fig.2.24, for Al 1100 - T0, Al 1100 - H14 and Al 5052 - H34 strips. The strips were rolled at 0.35, 10.4 and 3 and 4 rpm respectively. It is observed that both the reduction and roll pressure increase; almost linear relationships are evident. Further, the Al 1100 -H14 strips were lubricated with mineral oil, containing 1% oleic acid while the others were cleaned with freon. Contrary to expectations, friction was not found to be significantly lower for the lubricated strips.

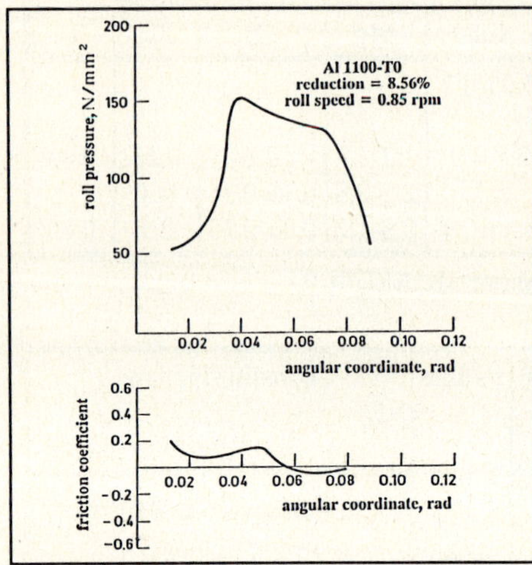

Fig.2.18 8.56 per cent reduction of 1100-T0 aluminium

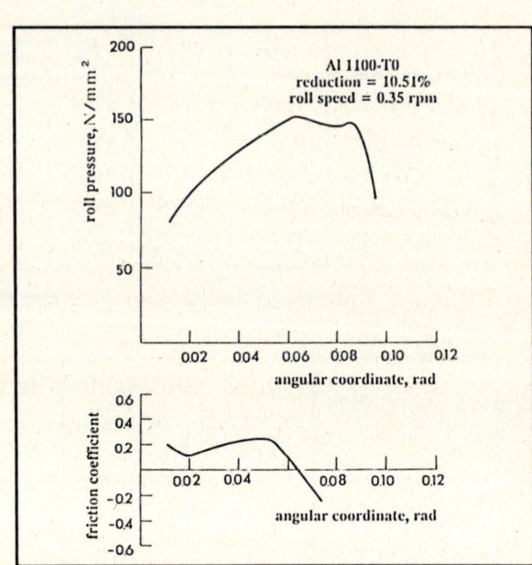

Fig.2.19 10.51 per cent reduction of 1100 -T0 aluminium

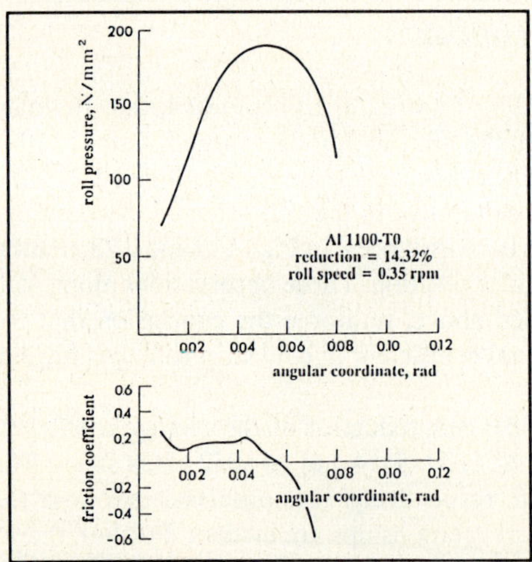

Fig.2.20 14.32 per cent reduction of 1100 -T0 aluminium

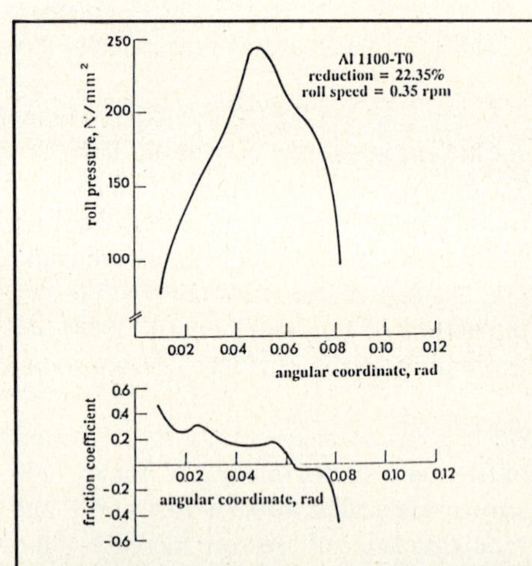

Fig.2.21 22.35 per cent reduction of 1100 -T0 aluminium

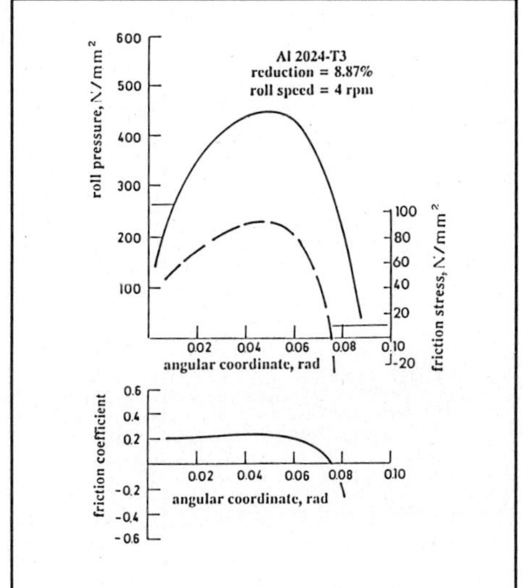

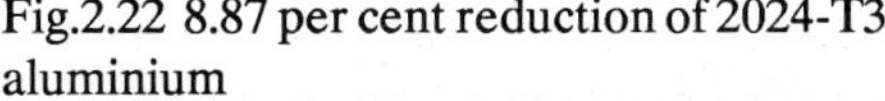

Fig.2.22 8.87 per cent reduction of 2024-T3 aluminium

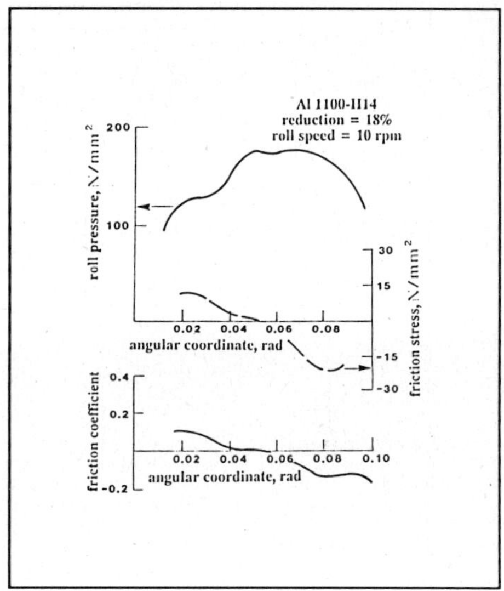

Fig.2.23 18 per cent reduction of 1100-H14 aluminium

One of the most important process parameters affecting the frictional conditions is expected to be the interfacial rubbing velocity. In order to determine its effect on μ the strip entry velocity for various rolling speeds and reductions was measured. By mounting two photodiodes 50 mm apart near the entry and measuring the time between the two signals generated by the passage of the strip, the entry speed was obtained, combining in fact the effect of both roll rpm and reduction. The dependence of the frictional coefficient on the entry velocity is given in Fig.2.25. A similar kind of behaviour for the four materials is observed. At low entry speeds, i.e. low rpm and low reduction, friction appears to be high. Some decrease to an apparently minimum value occurs, at approximately 65 - 80 mm/s, beyond which μ begins to rise. This observation confirms the fact that frictional effects are strongly dependent on the relative velocity of the contacting bodies.

Effect of material properties. Rabinowicz (1965) lists the volume and surface properties which influence surface interactions. Of those listed, parameters of significance in cold rolling were considered in this project for their effect on the average coefficient of friction. These are: the initial yield strength, surface hardness, strain hardening sensitivity and planar anisotropy. The dependence of the average coefficient of friction on the initial yield strength is shown in

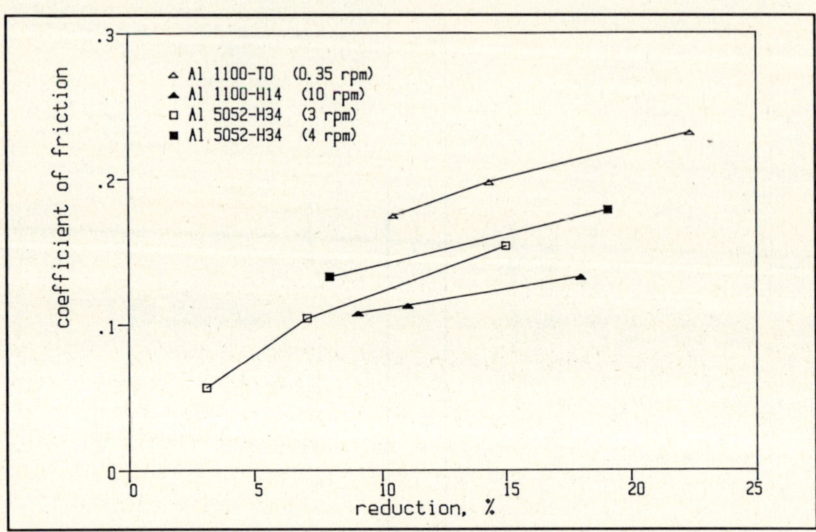

Fig.2.24 The effect of the reduction on the average friction coefficient.

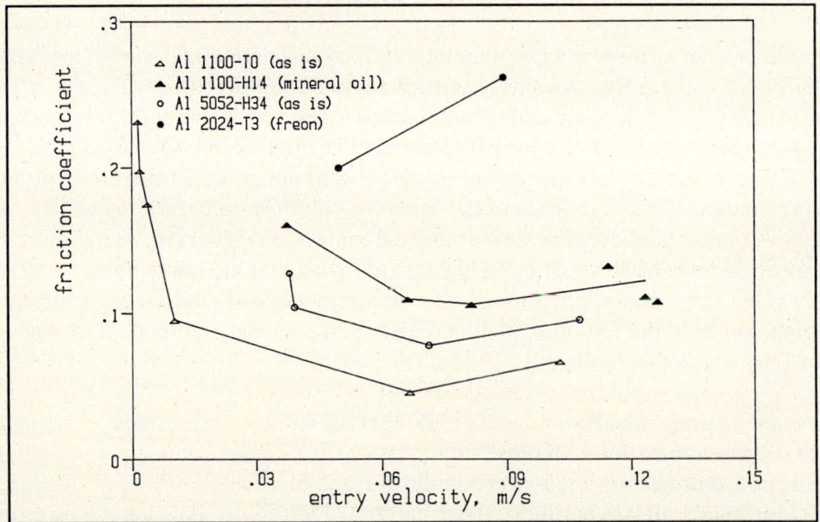

Fig.2.25 The effect of the entry velocity on the average friction coefficient.

Fig.2.26. A direct, linear relationship is observed. As expected, higher initial yield strength results in a higher coefficient of friction. Confirming this is the dependence of μ_{av} on surface hardness, also shown in Fig.2.26. Again, as expected, increased hardness leads to increased friction. Lower strain hardening sensitivity results in lower coefficients of friction, as seen in Fig.2.27. Finally, coefficient of friction appears to diminish with the growth of planar anisotropy - see again Fig.2.27.

Cold rolling with lubricants. Reduction of interfacial friction forces is of foremost importance in investigations of tool wear. In studies of friction in metalforming processes - and especially in cold, flat rolling - the objectives are somewhat different. Kraan's (1981) list is instructive; according to him the important performance characteristics of metalforming oils include low tool wear, good surface finish, easy removal of the lubricant, no corrosion and minimum health risks. Schey (1983) refers to his earlier paper (1967) in giving a list of the attributes of metalworking lubricants. He describes controlled friction, separation of surfaces, control of metal pickup, reduced wear, protection of surfaces, adaptability, compatibility with the die and the workpiece, rapid response, durability, controlled surface finish, insulation, cooling, stability, reactivity, harmless residues, application and removal, disposal, handling, safety and cost as properties to be considered. Neither synthetic nor natural lubricants can fulfill all or even most of the tasks given above. Additives are necessary which will enhance the lubricants' performance, as shown by Schey (1983). Boundary additives, determined by the deforming material, are used to reduce friction. Extreme pressure additives are needed to aid in avoiding lubricant breakdown. Solid additives may be necessary when high temperatures are encountered. Lubricants with additives are usually referred to as compound lubricants.

The pace of the development of synthetic metalworking fluids has increased significantly in the last two decades. Some of the reasons for this have been given by Sargent (1978) as the increasing strictness of environmental and health regulations, increasing costs and decreasing petroleum stocks. He further suggested that these difficulties may be overcome by the use of the synthetic lubricants, particularly those not based on petroleum products.

The questions to be answered are then these: can synthetic lubricants replace natural oils? Can they fulfill the needs of the metalforming industry and, in particular, the flat cold rolling process? An attempt, formulated to provide an answer to these questions, is described below. Using the experimental setup, described in Section 2.2.1 and presented also by Lenard (1987), roll pressures, interfacial shearing forces, roll separating forces, torques and forward slip were monitored during cold reduction of aluminium strips. First, the dependence of the coefficient of friction on the process parameters - reduction and speed of rolling - was established, using Freon spray to clean both rolls and both sides of the strip. Following that the experiments

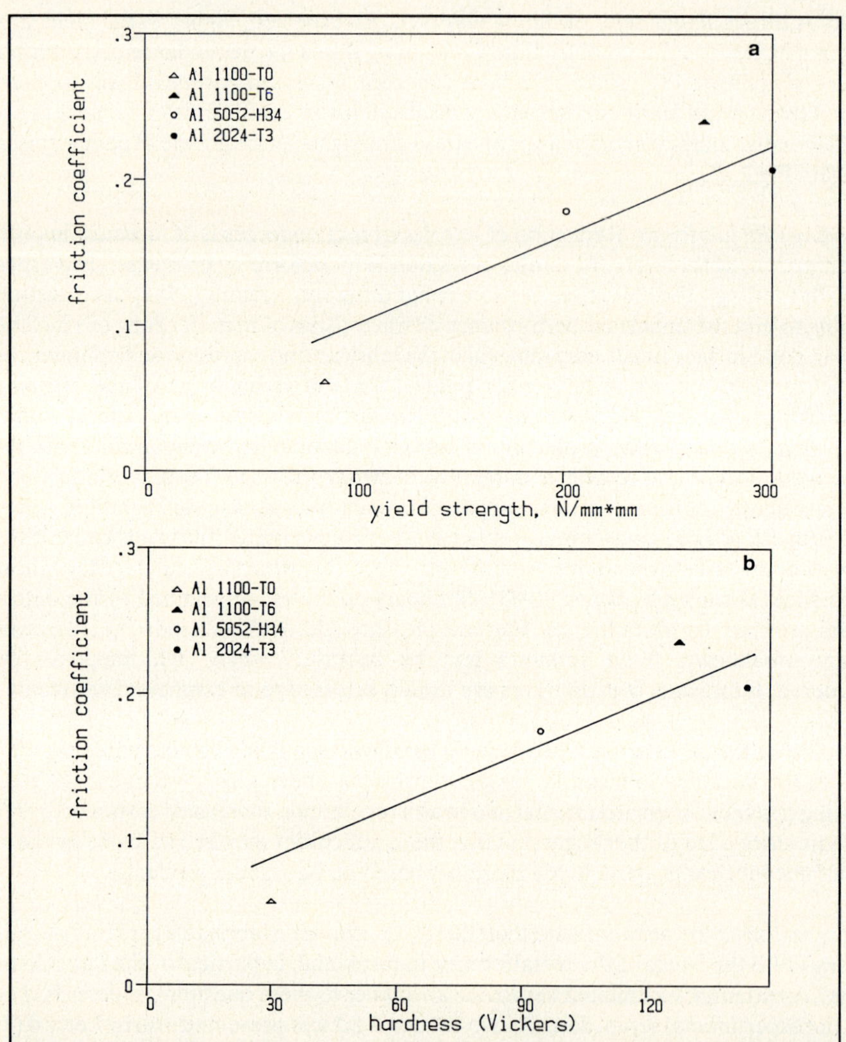

Fig.2.26 Influence of the yield strength (a) and hardness (b) on the coefficient of friction.

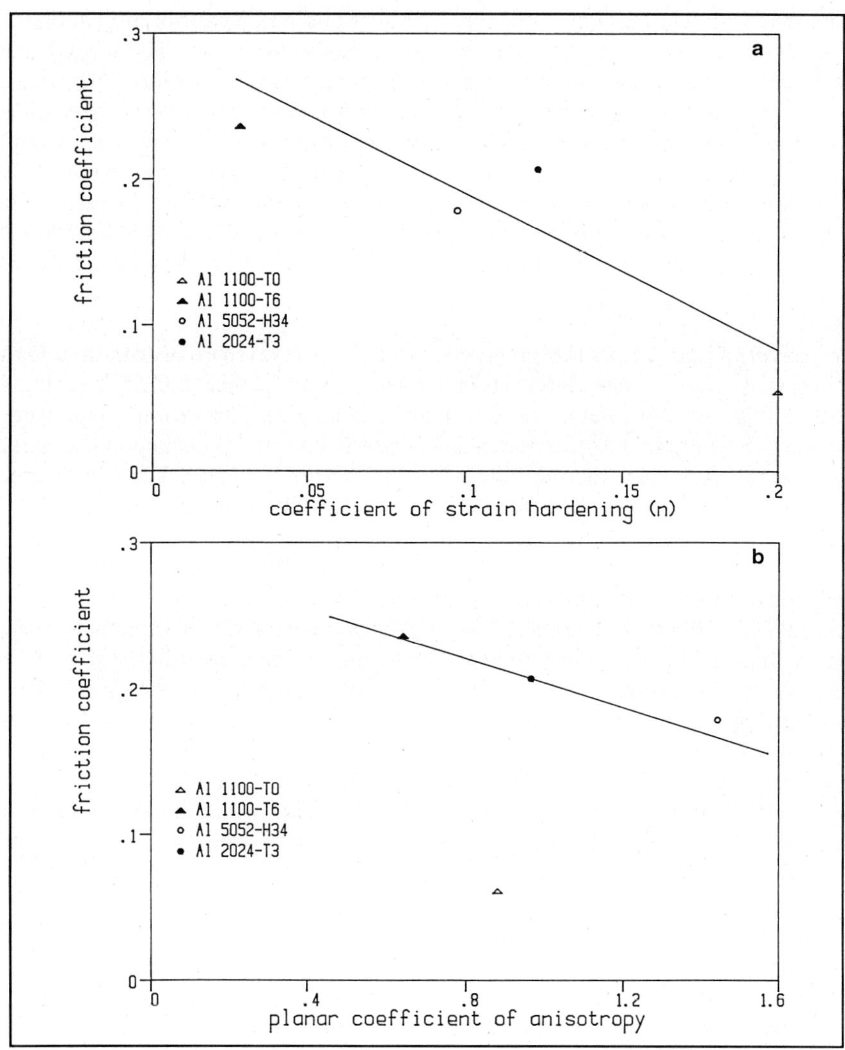

Fig.2.27 Influence of the coefficient of strain hardening (a) and planar coefficient of anisotropy (b) on the coefficient of friction.

were repeated with two types of lubricants. The first one was a natural oil, purchased in a drug store and mixed with 1% oleic acid, by volume, in the laboratory. The second lubricant was a synthetic polyalphaolefin containing two friction modifiers one of which was a fatty alcohol, derived from vegetable oil and the second was a fatty acid, derived from tallow. An antioxidant, extreme pressure and antiwear additives were also contained in the compound. Use of oils has been shown to lower forces and torques during flat rolling. Another question was then raised at this point - can difference in behaviour, resulting from the use of different oils be detected by the experimental equipment? Or, in other words, would metal forming oils of different viscosities cause measurable changes in the magnitudes of the reductions of the rolling loads?

Roll pressure and friction stress. Typical roll pressure and friction stress distributions in the roll gap during cold rolling of dry, cleaned aluminium strips are shown in Fig.2.28, using the Al 1100 - T0 alloy, usually referred to as commercially pure aluminium. Data from four experiments are given, each conducted at a roll speed of 3 rpm. The independent variable is chosen to be the reduction, varying from a low of 7.6% up to 31.8%. The curves appear as expected from previous work. The distributions of roll pressure are reasonably smooth. All have only one peak whose location coincides quite well with the location of the zero interfacial shear stress. As before, the reduction is observed to have a significant effect on the magnitude of the frictional forces and in turn on the coefficient of friction. As demonstrated by Karagiozis and Lenard (1985), higher reductions lead to higher average interfacial frictional coefficients. Adding the present results to those reported by Karagiozis and Lenard (1985) - see Fig.2.29 - confirms the earlier conclusion that the coefficient of friction is still a strong function of the rubbing speed and the reduction. Introduction of a mineral oil with 1% oleic acid as a lubricant into the roll gap causes significant changes in the distributions of the frictional stresses, as presented in Fig.2.30, where the interfacial normal and shear stresses are shown for approximately 8% reduction of the aluminium alloy. The neutral point must have been located very near the exit as it has not been found in either experiment - one conducted at 3 rpm while for the other the rolling speed was raised to 5.25 rpm - indicating the possible existence of hydrodynamic lubrication (Schey, 1983). The roll pressure distributions did not change in a pronounced manner.

Roll force and torque. Roll forces and torques measured during dry and lubricated experiments are shown plotted versus the reduction per pass in Figs 2.31 and 2.32 for an Al 1100 - T0 alloy and in Figs 2.33 and 2.34 for a somewhat harder Al 1100 - H14 metal. The lubricant used in the tests of Figs 2.31 and 2.32 was the mineral oil while the results of Figs 2.33 and 2.34 were obtained with the synthetic oil. It is observed that use of the lubricants reduced both roll forces

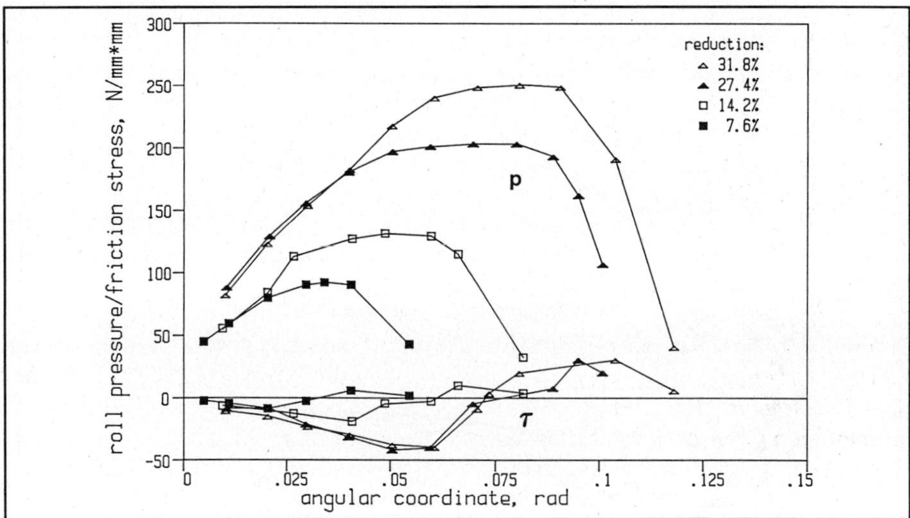

Fig.2.28 Roll pressure and friction stress distribution during dry rolling of aluminium.

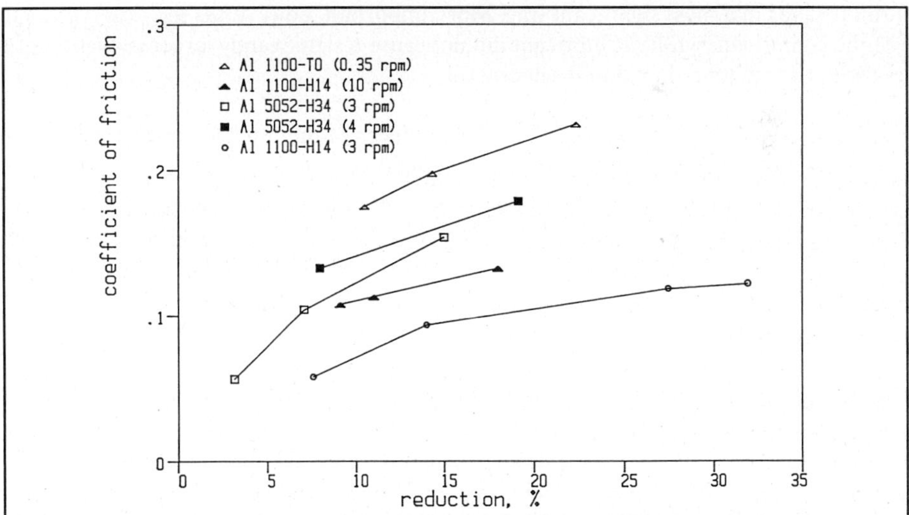

Fig.2.29 The effect of reduction on the average friction coefficient.

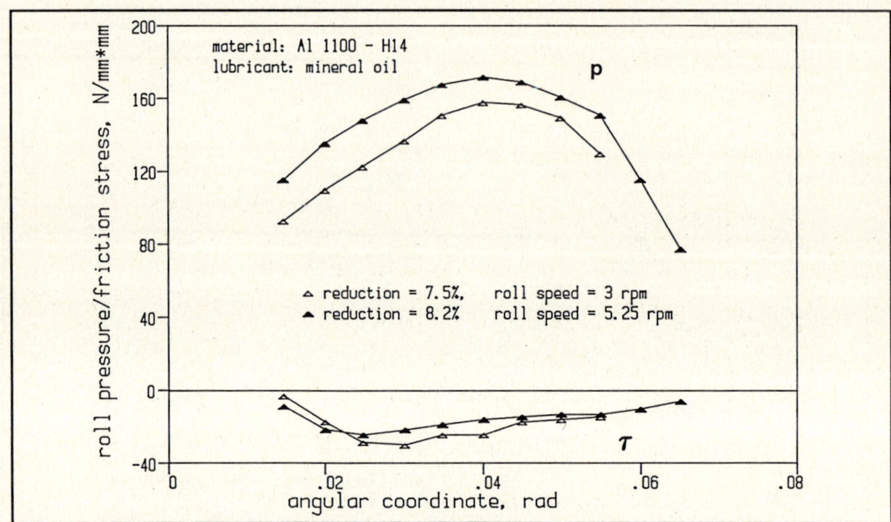

Fig.2.30 Roll pressure and friction stress distribution during rolling with a lubricant.

and roll torques in a most significant way. More important, however, is the observation that use of the compound, synthetic lubricant did not cause a significantly larger reduction of the forces and torques than the natural mineral oil.

Forward slip. Measuring the marks left by the scratches of the top roll on the strips allowed the determination of forward slip in the pass, defined as

$$s = \frac{l_2 - l_1}{l_1} \tag{2.11}$$

where l_1 is the distance between the scratchmarks on the roll and l_2 is the distance between the impressions on the strip. Results of measurements are given in Fig.2.35 where the forward slip values are presented in terms of the reduction per pass for unlubricated cold rolling and for data obtained during rolling with mineral oil as the lubricant. As well, the forward slip measurements obtained after rolling with the synthetic lubricant are also shown in this figure. The forward slip values rise with increasing reduction in all three cases, confirming the dependence of the frictional forces on reduction. Use of the mineral oil lubricant reduced the

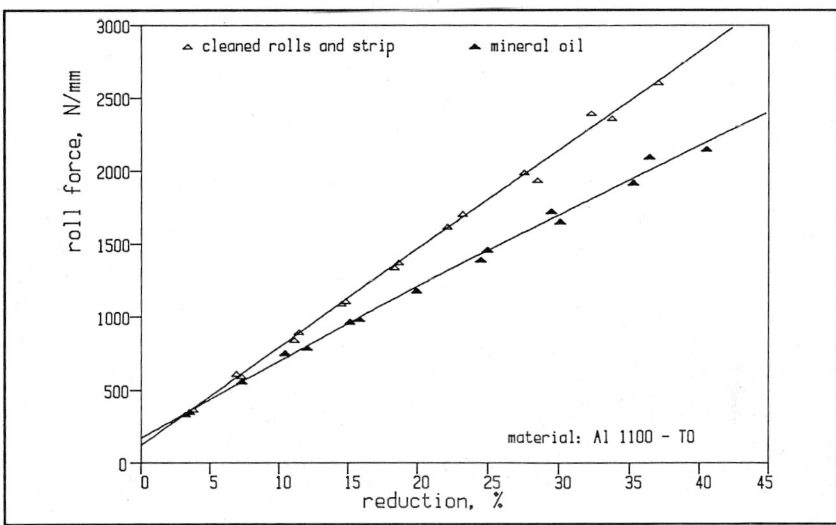

Fig.2.31 Roll force vs. reduction relationship for rolling of an Al 1100-T0 alloy.

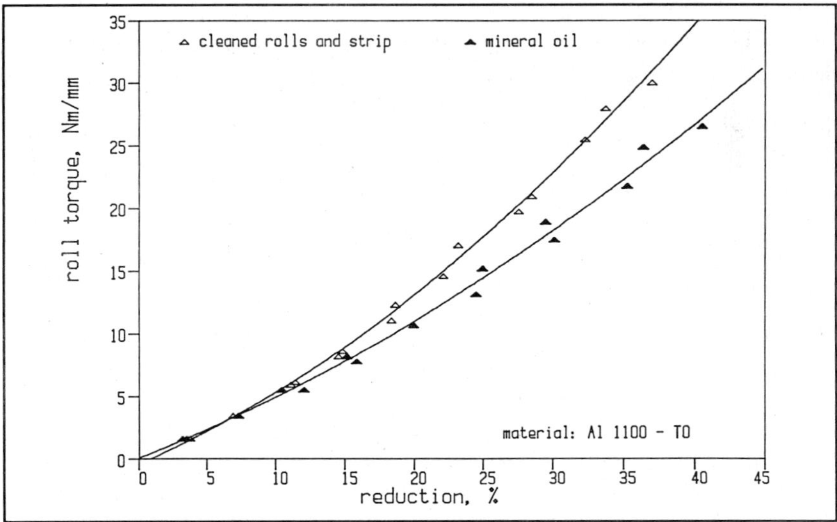

Fig.2.32 Roll torque vs. reduction relationship for rolling of an Al 1100-T0 alloy.

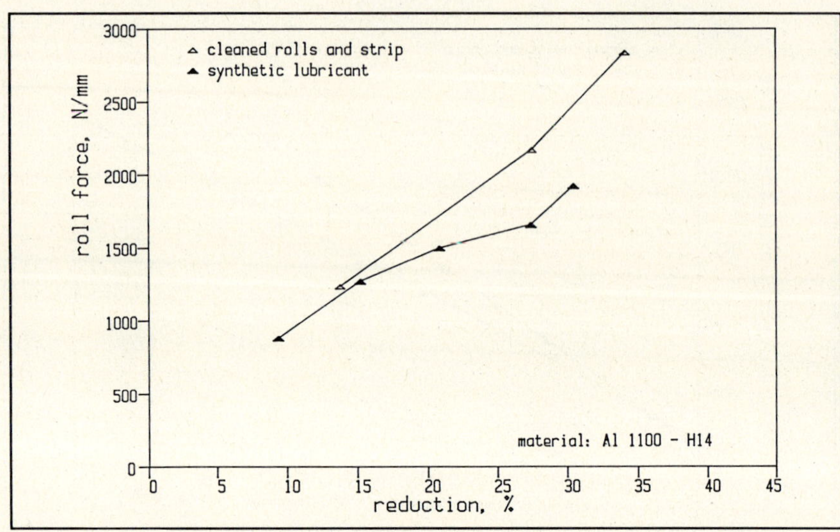

Fig.2.33 Roll force vs. reduction relationship for rolling of an Al 1100-H14 alloy.

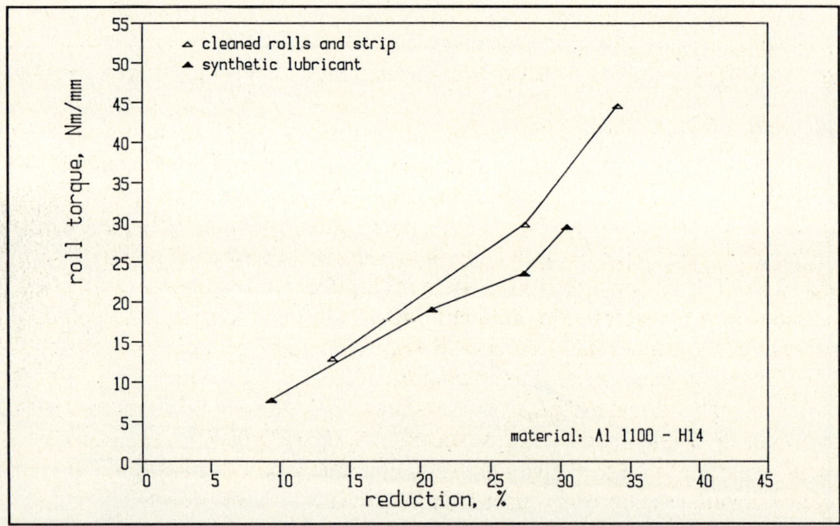

Fig.2.34 Roll torque vs. reduction relationship for rolling of an Al 1100-H14 alloy.

forward slip significantly. The synthetic lubricant also caused a very noticeable reduction of the forward slip, confirming the observations regarding the roll force and torque values of Figs 2.33 and 2.34.

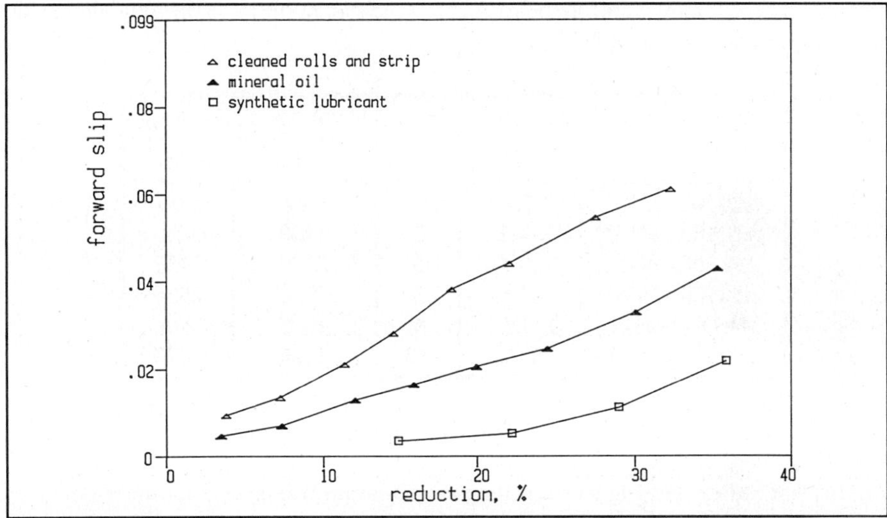

Fig.2.35 Forward slip vs. reduction relationship for rolling with various lubricants.

Substantiation of results. The accuracy of measurements obtained by the embedded pin technique has been questioned by Stephenson (1983), suggesting that a possible 50% error in frictional stresses may be unavoidable. It is undeniable that interrupting the contact surface, weakening the roll and thereby changing its mode of deformation, using pins of finite diameters, coping with the metal extruded into the clearance between the pins and their housing and accounting for frictional resistance there must affect the readings. Another drawback of the technique is its unlikely use in production-type mills making it impossible to substantiate experimental results under full scale conditions. However, since no other method capable of yielding directly the values of the interfacial forces in the roll gap during flat rolling is known to the authors - the indirect technique of Meierhofer and Stelson (1987) necessitates a certain amount of interpretation and the transducer described by Britten and Jeswiet (1986), while it looks promising, has yet to yield a significant amount of data - and since the measurements are needed, independent ways of substantiating the results have been sought.

Among these is the possibility of comparing the roll separating force measured by the transducers located under the bottom work roll to the integral of the pressure distribution, produced by the pin-technique. Selected results are given in Table 2.8. It is observed that the difference between the forces obtained by the two methods is less than 16%. Some confidence in the present results may therefore be restored. Note that material in the annealed condition is marked with a star in Table 2.7.

material	reduction %	rpm	roll force, N/mm	
			transducer	pins
Al 1100 - H14*	7.6	3	599	500
Al 1100 - H14*	14.2	3	1100	973
Al 1100 - H14*	27.4	3	2000	2138
Al 1100 - H14*	31.8	3	2394	2586
Al 1100 - H14	8.8	4	1006	886
Al 1100 - H14	11.2	10	1166	1070

Table 2.7 Comparison of forces measured by transducers and calculated by integration of measured pressure distribution.

Warm and hot rolling. During warm rolling the situation becomes even more complex. No more metal-to-metal contact is likely to occur. Instead, the interface may consist of several layers separating the rolled strip and the work roll, including oxide layers, water and steam. The locations and magnitudes of these layers depend on the process variables and are, quite probably, random in nature. Experimental results concerning the magnitude of the variation of the frictional coefficient in the roll gap during warm or hot rolling are rare. Bernick and Wandrei (1978) studied the effects of temperature, per cent reduction, entry thickness, steel grade and lubrication on the coefficients of friction in hot rolling of slabs and strips. The tests were carried out on a two-high reversing mill of 457 mm diameter work rolls and later substantiated on a six-stand hot strip mill. The frictional coefficients were determined by measuring the forward slip. The authors concluded that, at least within the range of parameters used, the coefficient of friction was independent of the temperature and the steel grade. It was however strongly dependent on the entry thickness and the shape coefficient defined as the ratio of the average strip thickness and the contact length. The values of μ reported by Bernick and Wandrei varied from a low of 0.16 corresponding to an entry thickness of 5 mm to about 0.39 for 25.4 mm. An anomaly, relating reduction to friction, was also reported. At high entry thicknesses, μ lowered as reduction was increased. At lower entry thicknesses the opposite tendency was observed.

Use of lubricants (a synthetic ester) decreased the coefficient of friction and the roll force more as the entry thickness increased. Roll wear also decreased significantly, raising the "equivalent tons rolled/roll change" by 27% - 43%.

The method of reflected caustics was applied by Theocaris et al. (1983) to measure the roll pressure distribution and the coefficient of friction. They used 89 mm diameter rolls made of plexiglass. The test material was lead, rolled at room temperature, to simulate rolling of steel at elevated temperatures. The measured frictional coefficients decreased almost linearly from entry to the no-slip point, beyond which it increased, again in a linear fashion. Theocaris et al. (1983) concluded that μ must be speed dependent.

Jarl (1988) used Ekelund's relationship as quoted by Wusatowski (1969), of the forward slip and the coefficient of friction in conjunction with numerical integration of the von Karman equation. The major conclusion of this study was that "...it is unreliable to evaluate the coefficient of friction in hot flat rolling by an application of a simple equation which presumes that the friction stress is unlimited and that the yield stress is constant in the roll gap".

Rowe (1977) considered the problems associated with lubrication during hot rolling of steel, copper and aluminium, stating that the demands on a hot rolling lubricant are not severe and that cooling of the work rolls is usually the prime objective. When steel is rolled the coolant is most often water, supplied in large quantities. When aluminium is being produced, difficulties may arise after the breakup of the oxide film, exposing the pure metal which may then adhere to the rolls. If no lubricant is used, the strip may in fact wrap itself around the roll completely. This in fact has happened in the writers' laboratory.

Information regarding specific values of the frictional coefficients in a hot rolling pass, for use in modelling, is practically nonexistent. Among the few formulae, giving μ as a function of roll material, rolling speed and temperature, are those presented by Geleji (1961):

- for steel rolls

$$\mu = 1.05 - 0.0005T - 0.056v$$

- for cast iron rolls

$$\mu = 0.94 - 0.0005T - 0.056v$$

- for ground steel or cast iron rolls

$$\mu = 0.82 - 0.0005T - 0.056v$$

were T is the temperature in ^{o}C and v is the rolling speed - which should probably be taken to mean the rubbing velocity.

Roberts (1974) and El-Kalay and Sparling (1968) have also studied hot friction. Reviewing the above leads to the same conclusion, arrived at when considering experimental studies of friction at room temperatures. What appears to be missing at this point is again a sufficiently large body of experimental data on the dependence of the frictional forces in hot strip rolling on process variables and material properties. The results of two experiments on friction in warm rolling of aluminium are presented below.

Friction in warm, dry rolling. Two experiments were conducted, one at a temperature of 210°C and the other at 155°C, the objectives of which were to begin a detailed study of the effect of temperature on the nature of dry, unlubricated friction during flat rolling of aluminium slabs. Two type K thermocouples were embedded in the tail end of each slab and the variations of temperature during the pass were monitored in addition to the roll force, torque and embedded transducer data. The procedure was unchanged except to take care that the slabs' temperatures during forward and reverse rolling were identical. The results are presented in Fig.2.36 in the form of roll pressure and interfacial friction stress plots. As expected, the frictional forces were greatly affected by the temperatures, higher values leading to higher frictional resistance.

2.2.3 Discussion

Some general conclusions may be drawn at this stage. It is concluded that the ratio of the interfacial shearing stress to normal pressure does not remain constant in the roll gap during warm and cold rolling of aluminium strips. As well, in most cases the location of maximum pressure is close to that of zero interfacial shear. In most cases smooth pressure distribution curves are found having one maximum. Two peaks are observed only in a few cases.

Some differences in the results are noticed when either uncleaned or cleaned strips are rolled. In both instances clearly defined neutral points are found. When lubricant is introduced however, the experimental results change markedly. No neutral point - i.e. no zero friction stress - is located, no doubt due to the different circumstances in the roll gap. The above results contradict some of the findings of van Rooyen and Backofen (1957). In the only other experiment on friction with lubrication reported in the literature, they presented large variations of the frictional coefficient. The current results show no such variation.

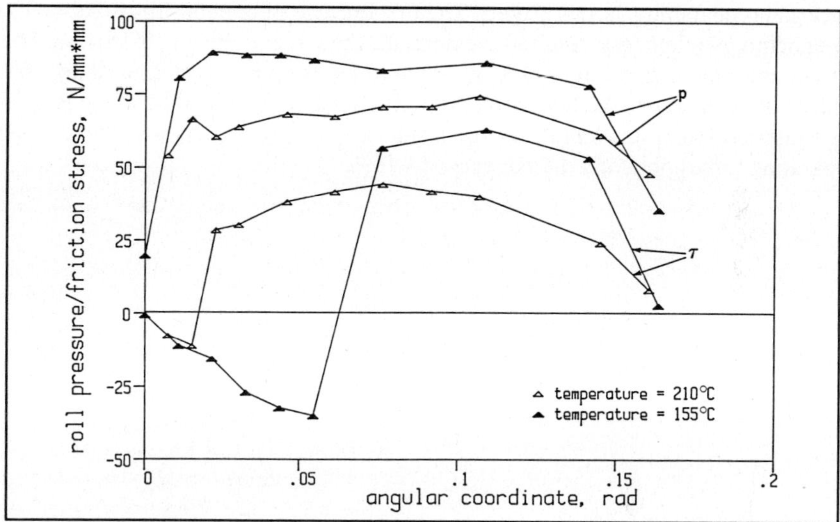

Fig.2.36 Roll pressure and friction stress distributions in warm rolling.

The locations of the measuring pins appear to have a major effect on the results. It is reasonable to expect that because of roll bending, especially with the uncrowned rolls used in this study, roll pressures are highest at the strip's longitudinal centre line and that they decline toward the edges. This expectation is confirmed by the present results. Roll pressures measured by a transducer with pins at the roll centre, see Lim and Lenard (1984), are significantly higher than those measured by the transducer with pins located side by side.

The most significant parameter affecting the friction conditions is found to be the rolling speed and the entry velocity. In general, the coefficient of friction decreases as the rolling speed increases. A minimum μ is indicated when its variation with the entry velocity is observed. Beyond that value, approximately 65 - 80 mm/s, the frictional coefficient increases with increasing entry speed.

As stated by van Rooyen and Backofen (1957) the results of embedded pin experiments are extremely sensitive to calibration errors and slight inaccuracies. They found that a 1% error in the reading of one of their transducers may cause a 10% error in the value of the frictional coefficient. In light of these comments, two separate ways were used to test for accuracy and repeatability in this project. The first one was obtained with Al 2024 - T3 strips, where eight independent tests resulted in essentially the same coefficient of friction. The other concerned the roll separating force. By comparing the values measured by the force transducers located under the lower rolls with the numbers obtained by integrating the roll pressure distribution, it was found that the results are, in fact, reasonable. The difference of the separating forces obtained did not exceed ± 16%.

CHAPTER 3

ONE-DIMENSIONAL MODELS OF FLAT ROLLING

There are three sections in this chapter. The first includes the description of the conventional mathematical models applicable for cold and hot rolling, developed by Orowan, Sims, Bland and Ford, Ford and Alexander and Tselikov. In the second a refinement of the one-dimensional models is presented and a new philosophy guiding the improvement of computational accuracy and consistency is described. The third section contains an assessment of the assumptions introduced in various models and experimental substantiation.

3.1 Conventional Models

3.1.1 Orowan's Theory

Derivation of basic equations. One of the most comprehensive one dimensional studies of hot and cold flat rolling is due to Orowan (1943). The equations that he developed do not appear to yield analytical solutions for the roll pressure. Researchers - see Bland and Ford (1948), Sims (1954), Ford and Alexander (1963) and Tselikov (1939) - have therefore devised various approximate solutions. Alexander (1972) reviewed and critically evaluated some of these approximations. As well, he presented an essentially complete solution of Orowan's equations using a fourth order Runge-Kutta technique to integrate the differential equations. In this section Alexander's paper will be reviewed.

The equation of equilibrium in the longitudinal direction, of the plastically deforming material in the roll gap, is derived by equating the sum of forces in that direction to zero. Considering Fig.3.1 this yields

$$-(\sigma_x + d\sigma_x)(h + dh) + \sigma_x h \pm 2\tau R' \cos\phi \, d\phi + 2pR' \sin\phi \, d\phi = 0 \tag{3.1}$$

Neglecting terms that are at least an order of magnitude less than the others leads to

$$\frac{d}{d\phi}(\sigma_x h) = 2R'(p\sin\phi \pm \tau\cos\phi) \qquad (3.2)$$

where the upper algebraic sign refers to the region between the neutral point and exit and the lower sign denotes the region between the neutral point and entry. In equations (3.1) and (3.2) σ_x is the stress in the x direction, h is the current thickness of the slab, R' is the radius of the deformed roll, p is the roll pressure and τ designates the shear stress between the roll and the slab. It is the roll pressure that must be determined from equation (3.2). Several other equations, relating the other variables to the angular coordinate ϕ must therefore be derived. One is the variation of thickness in the roll gap which can be shown to be

$$h = h_2 + 2R'(1 - \cos\phi) \qquad (3.3)$$

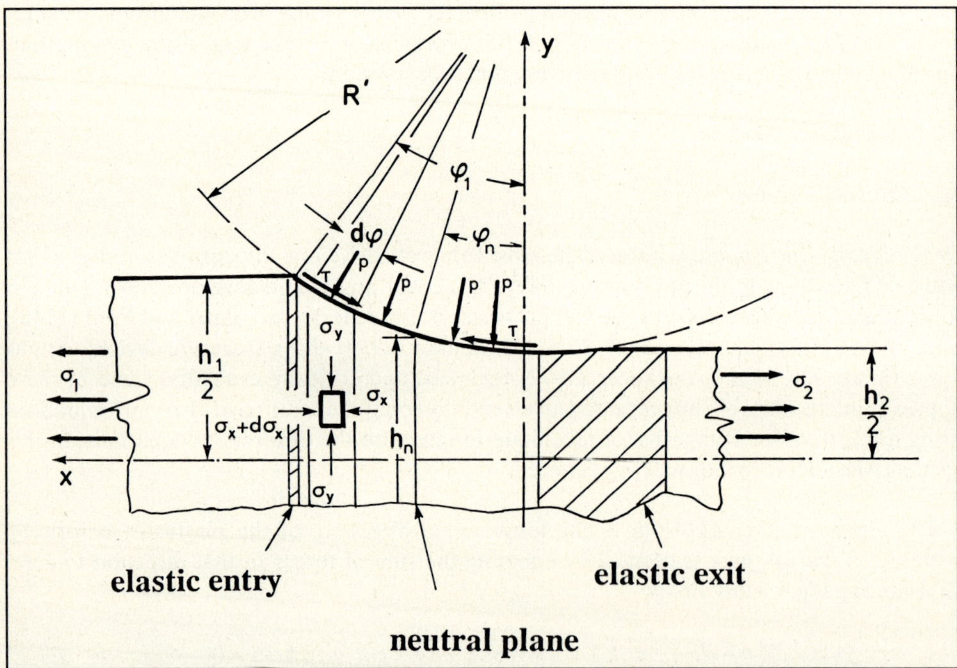

Fig.3.1 Forces and stresses acting upon an elemental section of the strip between the rolls.

The stress in the x direction is related to the roll pressure and this relationship is obtained from the Huber-Mises criterion of plastic flow. The horizontal and vertical stress components are related by

$$\sigma_y - \sigma_x = 2k \tag{3.4}$$

Letting σ_y represent the resultant stress caused by the interfacial forces

$$\sigma_y = p \mp \tau \tan\phi \tag{3.5}$$

the required equation is obtained as

$$\sigma_x = p - 2k \mp \tau \tan\phi \tag{3.6}$$

Substituting equation (3.6) into (3.2) gives the equilibrium equation, as presented by Alexander

$$\frac{d}{d\phi}[h(p - 2k \mp \tau \tan\phi)] = 2R'(p \sin\phi \pm \tau \cos\phi) \tag{3.7}$$

Alexander considers two possible cases of friction between the roll and the strip. He states that either the traditional Coulomb friction assumption given by equation (1.7) is valid; or in the extreme case, if sticking is encountered, equation (1.9) should be employed. For the numerical solution the smaller of the frictional conditions is used. Substituting equations (3.6) and (1.7) or (1.9) in (3.2) yields the equations of equilibrium for the two cases of friction

$$\frac{dp}{d\phi} = g_1(\phi)p + g_2(\phi) \qquad \text{for:} \qquad \tau = \mu p \tag{3.8}$$

where

$$g_1(\phi) = \pm\mu \sec\phi \frac{\frac{2R'}{h} + \sec\phi}{1 \mp \mu \tan\phi}$$

$$g_2(\phi) = \frac{\frac{2R'}{h} 2k \sin\phi + \frac{d(2k)}{d\phi}}{1 \mp \mu \tan\phi}$$

and

$$\frac{dp}{d\phi} = g_3(\phi) \qquad for: \qquad \tau = k \qquad (3.9)$$

where

$$g_3(\phi) = 2k\left[\frac{R'}{h}\sin\phi(2 \pm \tan\phi) \pm \left(\frac{R'}{h}\cos\phi + \frac{1}{2}\sec^2\phi\right)\right] + \frac{dk}{d\phi}(2 \pm \tan\phi)$$

For an analytical or a numerical solution of the problem, equation (3.3) describing the dependence of the strip thickness on the deformed roll radius, exit thickness and angular coordinate, is also required. Boundary conditions must be prescribed; at entry the roll pressure is given by

$$p_1 = 2k_1 - \sigma_1 - \tau \tan\phi_1 \qquad (3.10)$$

and at exit

$$p_2 = 2k_2 - \sigma_2 \qquad (3.11)$$

In equations (3.10) and (3.11) the subscripts 1 and 2 refer to conditions at entry and exit respectively, while σ_1 and σ_2 designate the front and back tensions.

Two further parameters must still be defined. These include the material properties of the rolled stock and the manner of the elastic deformation of the work roll.

Material properties. In equations (3.8) and (3.9) the parameter k is defined by Alexander (1972) as the yield strength of the material in pure shear. It is related to the uniaxial yield stress in plane strain compression by equation (1.6). Alexander (1972) relates the yield stress in shear to the compressive strain of the strip in the roll gap using the formula

$$2k = \frac{2}{\sqrt{3}}\sigma_{po}\left(1 + \frac{2}{\sqrt{3}}B\ln\frac{h_0}{h}\right)^n \tag{3.12}$$

where σ_{po} is the initial, uniaxial yield stress of the material prior to any deformation, h_0 is the thickness at which the strip was in its last annealed state and h is the current thickness of the strip. B and n are material constants which are the strength parameter and strain hardening coefficient, respectively.

Roll deformation. During rolling the originally cylindrical rolls deform in a very complex manner. Substantial, though still elastic, distortion of the circular cross section occurs in addition to significant bending of the whole roll. This deformation implies that the shape of the roll gap will not be geometrically well defined. It will, in fact, depend on roll elasticity, roll pressure and friction distribution, rotational speed of the rolls, temperature of the rolled stock, to mention just a few factors.

The traditional method to determine the deformed roll shape is due to Hitchcock (1935) who assumed the existence of an elliptical pressure distribution and calculated the radius of the deformed roll, R'. It must be noted that R' implies an assumption that the roll shape, after deformation, remains circular. The new radius of curvature is larger than the original, however. Mathematically, the roll force per unit width F, original roll radius R, deformed roll radius R' and the draft Δh are related by

$$\frac{R'}{R} = 1 + \frac{CF}{\Delta h} \tag{3.13}$$

where

$$C = \frac{8(1 - \nu^2)}{\pi E}$$

The constant C in equation (2.13), calculated for various roll materials, is:

- for cast iron rolls	C = 3.4 - 4.8 $10^{-5} mm^2/N$
- for steel rolls	C = 2.1 - 2.5 $10^{-5} mm^2/N$
- for sintered carbide rolls	C = 0.8 - 1.0 $10^{-5} mm^2/N$

Equation (3.13) is known as Hitchcock's formula. Ford et al. (1951) modified this equation to include the effects of elastic entry and exit, that is to account for the fact that the magnitude of roll gap is affected by the elasticity of the rolled stock (Fig.3.1). The modified equation used by Alexander (1972) is

$$\frac{R'}{R} = 1 + \frac{CF}{(\sqrt{\Delta h + \Delta h_e + \Delta h_t} + \sqrt{\Delta h_e})^2} \tag{3.14}$$

where

$$\Delta h_e = (1 - \nu^2)(2k_2 - \sigma_{e2})\frac{h_2}{E}$$

$$\Delta h_t = (h_2\sigma_2 - h_1\sigma_1)\frac{\nu(1+\nu)}{E}$$

Finally, the contributions of elastic entry and elastic exit to the separating force are given by

$$F_{e1} = \frac{(1-\nu^2)h_1}{4}\sqrt{\frac{R'}{\Delta h}\frac{(2k_1 - \sigma_{e1})^2}{E}} \tag{3.15}$$

$$F_{e2} = \frac{2}{3}\sqrt{\frac{R' h_2 (1-\nu)^2}{E}}(2k_2 - \sigma_{e2})^{1.5} \tag{3.16}$$

Equations (3.15) and (3.16) must, of course, be solved by an iterative technique. Here σ_{e1} and σ_{e2} represent front and back tensions respectively, as affected by elastic entry and exit and are given by

$$\sigma_{e1} = \sigma_1 - 2\mu\frac{F_{e1}}{h_1} \tag{3.17}$$

$$\sigma_{e2} = \sigma_2 - 2\mu\frac{F_{e2}}{h_2} \tag{3.18}$$

where μ is the friction coefficient and h_1 and h_2 represent the initial and final thickness of the strip, respectively.

Roll force and roll torque. After equations (3.8) or (3.9) have been integrated for the roll pressure it is then possible to evaluate the roll separating force and the roll torque. Alexander gives the following formulae

$$F = R'\int_0^{\phi_1} p\cos\left(\phi - \frac{\phi_1}{2}\right)d\phi + R'\left[\int_{\phi_n}^{\phi_1} \tau\sin\left(\phi - \frac{\phi_1}{2}\right)d\phi - \int_0^{\phi_n} \tau\sin\left(\phi - \frac{\phi_1}{2}\right)d\phi\right] \tag{3.19}$$

$$M = R'(R'-R)\int_0^{\phi_1} p\sin\left(\phi - \frac{\phi_1}{2}\right)d\phi + R'\int_{\phi_n}^{\phi_1}\left[R'\tau - (R'-R)\tau\cos\left(\phi - \frac{\phi_1}{2}\right)\right]d\phi - R'\int_0^{\phi_n}\left[R'\tau - (R'-R)\tau\cos\left(\phi - \frac{\phi_1}{2}\right)\right]d\phi \tag{3.20}$$

In equations (3.19) and (3.20) ϕ_n represents the angular coordinate of the neutral point and ϕ_1 is the angle of bite.

3.1.2 Sims' Method

The theory presented by Sims (1954) is reviewed in the present section. It is equation (3.7) that must be solved first for the roll pressure in order to allow for the estimation of separating forces and roll torque. There are several logical steps that may be taken to reduce the complexity of the equation. Sims (1954) assumes in his paper that as the roll gap angle ϕ_1 would in most cases be a small angle,

$$\tan\phi = \sin\phi = \phi \quad ; \quad \cos\phi = 1 \quad ; \quad 1 - \cos\phi = \frac{1}{2}\phi^2 \tag{3.21}$$

Introducing the correction for nonuniform deformation he obtains

$$\frac{d}{d\phi}\left[h\left(p - \frac{\pi}{2}k \mp \tau\phi\right)\right] = 2R'(\phi p \pm \tau) \tag{3.22}$$

As well the strip thickness in the roll gap may be approximated by

$$h = h_2 + R'\phi^2 \tag{3.23}$$

Sims further assumes that the product $\tau\phi$ is negligibly small in comparison to the other terms. Hence (3.22) becomes

$$\frac{d}{d\phi}\left[h\left(p - \frac{\pi}{2}k\right)\right] = 2R'(p\phi \pm \tau)$$

Allowing the interfacial friction stress to remain constant and equal to the yield strength of the material in shear, according to (1.9), Sims' equation results in

$$\frac{d}{d\phi}\left[h\left(p - \frac{\pi}{2}k\right)\right] = 2R'(p\phi \pm k) \tag{3.24}$$

which forms the basis of his theory of rolling. Using the present notation, Sims' solution for the roll pressure is given, commencing at the exit, as

$$\frac{p}{2k} = \frac{\pi}{4}\ln\frac{h}{h_1} + \frac{\pi}{4} + \sqrt{\frac{R'}{h_2}}\tan^{-1}\left(\sqrt{\frac{R'}{h_2}}\phi\right) \tag{3.25}$$

and for the portion of the roll gap from entry to the neutral point

$$\frac{p}{2k} = \frac{\pi}{4}\ln\frac{h}{h_1} + \frac{\pi}{4} + \sqrt{\frac{R'}{h_2}}\tan^{-1}\left(\sqrt{\frac{R'}{h_2}}\phi_1\right) - \sqrt{\frac{R'}{h_2}}\tan^{-1}\left(\sqrt{\frac{R'}{h_2}}\phi\right) \tag{3.26}$$

The location of the neutral point ϕ_n may be calculated from

$$\frac{\pi}{4}\ln(1-r) = 2\sqrt{\frac{R'}{h_2}}\tan^{-1}\left(\sqrt{\frac{R'}{h_2}}\phi_n\right) - \sqrt{\frac{R'}{h_2}}\tan^{-1}\left(\sqrt{\frac{r}{1-r}}\right) \tag{3.27}$$

The roll separating force per unit width is given by Sims (1954) as

$$F = 2kR'\left[\frac{\pi}{2}\sqrt{\frac{h_2}{R'}}\tan^{-1}\sqrt{\frac{r}{1-r}} - \frac{\pi\phi_1}{4} - \ln\frac{h(\phi_n)}{h_2} + \frac{1}{2}\ln\frac{h_1}{h_2}\right] \tag{3.28}$$

and the roll torque per unit width for one roll as

$$M = 2kRR'\left(\frac{\phi_1}{2} - \phi_n\right) \tag{3.29}$$

In equation (3.28) the term $h(\phi_n)$ designates the strip thickness at the neutral point.

3.1.3 Bland and Ford's Technique

In addition to the assumptions made by Orowan (1943), Bland and Ford (1948) further assumed that the roll pressure is equal to the stress in the vertical direction σ_y. This allowed a closed form solution for the roll pressure to be obtained. On the exit side of the neutral point then

$$\frac{p}{2k} = \frac{h}{h_2}\left(1 - \frac{\sigma_2}{2k_2}\right)\exp(\mu H) \tag{3.30}$$

and on the entry side

$$\frac{p}{2k} = \frac{h}{h_1}\left(1 - \frac{\sigma_1}{2k_1}\right)\exp[\mu(H_1 - H)] \tag{3.31}$$

In these relations σ_1 and σ_2 represent interstand tensions applied at entry and exit respectively, and H is given by

$$H = 2\sqrt{\frac{R'}{h_2}}\tan^{-1}\left(\sqrt{\frac{R'}{h_2}}\phi\right) \tag{3.32}$$

Assuming an average, constant yield strength in shear, $\bar{k}$, in the pass, the roll separating force and the roll torque are obtained as

$$F = R'2\bar{k}\left\{\int_0^{\phi_n}\frac{h}{h_2}\exp(\mu H)d\phi + \int_{\phi_n}^{\phi_1}\frac{h}{h_1}\exp[\mu(H_1 - H)]d\phi\right\} \tag{3.33}$$

and

$$M = RR'2\bar{k}\left\{\int_0^{\phi_n}\frac{h}{h_2}\exp(\mu H)\phi d\phi + \int_{\phi_n}^{\phi_1}\frac{h}{h_1}\exp[\mu(H_1 - H)]\phi d\phi\right\} \tag{3.34}$$

The mean yield strength, $\bar{k}$, is given by

$$\bar{k} = \frac{1}{\phi_1}\int_0^{\phi_1} k\, d\phi \tag{3.35}$$

and the neutral angle is calculated from

$$\phi_n = \sqrt{\frac{h_2}{R'}}\tan\sqrt{\frac{h_2}{R'}\frac{H_n}{2}} \tag{3.36}$$

with

$$H_n = \frac{H_1}{2} - \frac{1}{2\mu}\ln\left[\frac{h_1}{h_2}\left(\frac{1-\frac{\sigma_2}{2k_2}}{1-\frac{\sigma_1}{2k_1}}\right)\right] \tag{3.37}$$

In equations (3.31), (3.33) and (3.37) H_1 represents the value of H calculated from equation (3.32) for $\phi = \phi_1$.

3.1.4 Ford and Alexander's Method

In a paper published in 1963, Ford and Alexander presented two formulae for the calculation of roll separating forces and roll torques in hot, flat rolling. They are

$$\frac{F}{\bar{k}h_1} = 1.571\left(\sqrt{\frac{Rr}{h_1}} + \frac{R}{h_1}\frac{r}{2-r}\right) \tag{3.38}$$

and

$$\frac{M}{\bar{k}h_1^2}=\frac{Rr}{h_1}\left[1.571+(2/3)\frac{3-2r}{(2-r)^2}\sqrt{\frac{Rr}{h_1}}\right] \tag{3.39}$$

where F represents the separating force per unit width, M is the roll torque per unit width for both rolls, $\bar{k}$ is the average shear yield strength in the pass, R is the roll radius, h_1 is the strip thickness at entry and r represents the reduction.

3.1.5 Tselikov's Solution

A slightly different version of the solution of the equilibrium equation was developed by Tselikov (1939) presented by Tselikov and Grishkov (1970), published also by Morawiecki et al. (1986). Tselikov's assumptions which are introduced into the solution include the substitution of the arc of contact by its chord, Coulomb's model of friction and a constant value of the yield strength. The equilibrium equation written in Cartesian coordinates is

$$\frac{dp}{dx}-\frac{2k}{y}\frac{dy}{dx}\pm\frac{\tau}{y}=0 \tag{3.40}$$

A general integral for equation (3.40) is given by Tselikov as

$$p=\exp\left(\mp\int\frac{\mu}{y}dx\right)\left[C+\int\frac{2k}{y}\exp\left(\mp\int\frac{\mu}{y}dx\right)dy\right] \tag{3.41}$$

Introduction of the equation of the chord

$$y=0.5\left(h_2+\frac{h_1 r}{l_d}x\right) \tag{3.42}$$

into the relationship (3.41) yields

$$p=\exp\left(\mp\int\frac{\delta}{y}dy\right)\left[C+\int\frac{2k}{y}\exp\left(\mp\int\frac{\delta}{y}dy\right)dy\right] \tag{3.43}$$

where

$$\delta = \frac{2\mu l_d}{h_1 r} \quad ; \qquad l_d = \sqrt{R h_1 r}$$

and after integration

$$p = y^{\mp\delta} C \pm \frac{2k}{\delta} \tag{3.44}$$

The integration constant C is determined from the boundary conditions which, in a general case, include

$$p = \xi_1 2k_1 \qquad \text{at entry}$$

$$p = \xi_2 2k_2 \qquad \text{at exit}$$

where ξ_1 and ξ_2 are the tension coefficients defined as

$$\xi_1 = \frac{2k_1 - \sigma_1}{2k_1}$$

$$\xi_2 = \frac{2k_2 - \sigma_2}{2k_2}$$

and σ_1 and σ_2 are the stresses of the back and front tensions, respectively.

Introduction of the boundary conditions into equation (3.43) yields the relationship which describes the roll pressure distribution along the arc of contact

$$p = 2\frac{k}{\delta}\left[\left(\frac{h}{h_2}\right)^{\delta}(\xi_2\delta + 1) - 1\right] \qquad \text{for the forward slip zone} \tag{3.45}$$

$$p = 2\frac{k}{\delta}\left[\left(\frac{h_1}{h}\right)^{\delta}(\xi_1\delta - 1) + 1\right] \qquad \text{for the backward slip zone} \tag{3.46}$$

In equations (3.45) and (3.46) k is the yield strength in shear, h_1, h_2 are the entry and exit thicknesses respectively, and h represents the current thickness of the strip.

Tselikov assumes that in the hot rolling process both tensions are equal to zero and the yield strength is constant in the deformation zone. Integration of equations (3.45) and (3.46) with these assumptions yields the final formula for the roll separating force per unit width in the hot rolling process

$$F = \frac{2kl_d}{h_1 r}\frac{h_n}{\delta - 1}\left[\left(\frac{h_n}{h_1}\right)^{\delta} - 1\right] \tag{3.47}$$

In equation (3.47) h_n represents the strip thickness at the neutral plane. This thickness is determined from the condition that at the neutral plane the roll pressures calculated for the forward slip zone and backward slip zone from equations (3.45) and (3.46) respectively, are equal. Introduction of this condition yields

$$h_n = h_2\left[\frac{1 + \sqrt{1 + (\delta^2 + 1)\left(\frac{h_1}{h_2}\right)^{\delta}}}{\delta + 1}\right]^{\frac{1}{\delta}} \tag{3.48}$$

Analyzing the cold rolling process, Tselikov introduces tensions and assumes that the yield strength in shear is a constant, equal to k_1 in the backward slip zone and to k_2 in the forward slip zone. Integration of equations (3.45) and (3.46) with the above assumptions gives

$$F = \frac{4k_1\xi_1}{\mu h_1(2-r)}\left[\left(\frac{k_1\xi_1}{k_2\xi_2}\right)^{a_0}\exp\left(\frac{2\mu l_d}{h_1(2-r)}\right) - \frac{k_1\xi_1}{k_2\xi_2}a_0 - \frac{h_2}{h_1 + h_2}\right] \tag{3.49}$$

where

$$a_0 = \frac{h_1}{h_1 + h_2}$$

Tselikov's formula for cold rolling (3.49) is generally used together with Hitchcock's method of calculation of the elastic flattening of the roll. However, Tselikov and Grishkov (1970) have also presented their own method for the evaluation of the length of contact which accounts for the elastic deformation of both roll and the strip. (See also Jaglarz et al., 1980).

3.2 Refinement of the Conventional Models

In this section a refinement of the Orowan (1943) technique for flat rolling is described in detail - see Atreya and Lenard (1979), Saeed and Lenard (1980) and Roychoudhuri and Lenard (1984). The underlying philosophy guiding the development of this method of computation is to achieve maximum computational accuracy and consistency while using a minimum of arbitrary assumptions. The model consists of several parts. These include:

(a) plastic flow of the rolled strip in the roll gap,
(b) elastic compression and unloading of the strip at entry and exit,
(c) flattening of the work roll,
(d) frictional conditions at the roll-material interface,
(e) material's resistance to deformation.

The assumptions, limiting the validity of the technique, are listed below:

(a) no spread of the strip is considered,
(b) plane strain plastic flow is present,
(c) the compression of the strip is homogeneous,
(d) the rolled material is isotropic, homogeneous and in the plastic state, incompressible,
(e) the problem is quasistatic,
(f) there is no mill stretch.

3.2.1 Derivation of the basic equations

A schematic diagram of the flat rolling process is shown in Fig. 1.1 with the various regions of concern clearly indicated. These include the plastically flowing zone of the strip, bounded by the elastic compression and elastic recovery regions. As well, the deformed roll contour, whose exact shape is to be determined as part of these computations, is shown. The task is then described as the derivation of a mathematical model that describes the mechanics of the process, accounting for the pressure distribution on the walls of an arbitrarily shaped die whose shape itself depends on the pressure distribution. The technique therefore, has to be iterative.

Plastic region. The equation of equilibrium in the longitudinal direction is obtained as

$$\frac{d}{dx}\left[h\left(p-2k\mp\tau\frac{dy}{dx}\right)\right]=2\left(p\frac{dy}{dx}\pm\tau\right) \tag{3.50}$$

where h designates the current strip thickness, y is a coordinate with the origin at the centre of the upper roll, p is the roll pressure, k is the flow strength of the strip in shear and τ is the interfacial shear stress. The upper sign holds for the exit and the lower sign for the entry conditions. This convention will be adhered to in all subsequent developments.

The deformed roll shape is given by a function $y = f(x)$ which is determined as a part of the calculations. The strip thickness at an arbitrary location is expressed by

$$h=2(y+C) \tag{3.51}$$

where C is half of the centre to centre distance of the rolls.

The two frictional conditions at the strip-roll contact usually considered are represented by equations (1.7) and (1.9). If the original Amonton-Coulomb theory is assumed to hold, the coefficient of friction in equation (1.7) is taken to remain constant. If the adhesion theory of friction is assumed to be valid the frictional coefficient should be related to material and process parameters, as shown in Section 2.2 and in earlier publications by Lim and Lenard (1984) and Karagiozis and Lenard (1985) and a functional relationship

$$\mu = \mu \text{ (material and process parameters, independent variables)} \tag{3.52}$$

should be used. This relation may be an empirical equation or values of μ may be supplied directly in a digital form. Using the first frictional condition (1.7) as well as equation (3.51), equation (3.50) becomes

$$\frac{dp}{dx}=pg_1(x,y)+g_2(x,y) \tag{3.53}$$

where

$$g_1(x,y) = \pm \frac{\mu\left[\frac{d^2y}{dx^2}(y+C) + \left(\frac{dy}{dx}\right)^2 + 1\right]}{(y+C)\left(1 \mp \mu\frac{dy}{dx}\right)}$$

$$g_2(x,y) = \frac{2\frac{dk}{dx}(y+C) + 2k\frac{dy}{dx}}{(y+C)\left(1 \mp \mu\frac{dy}{dx}\right)}$$

while if equation (1.9) is valid the equation of equilibrium takes the form

$$\frac{dp}{dx} = k\frac{\left\{2\frac{dy}{dx} \pm \left[\left(\frac{dy}{dx}\right)^2 + 1\right]\right\}}{y+C} + 2\frac{dk}{dx}\left(1 \pm \frac{1}{2}\frac{dy}{dx}\right) \pm k\frac{d^2y}{dx^2} \tag{3.54}$$

The boundary conditions prescribe the roll pressure at entry to the plastic zone as

$$p_{p1} = 2k_1 - \sigma_{x1} - \mu p_1 \frac{dy}{dx} \tag{3.55}$$

and at exit from the plastic zone

$$p_{p2} = 2k_2 - \sigma_{x2} \tag{3.56}$$

In equations (3.55) and (3.56) σ_{x1} and σ_{x2} designate the stresses, at entry to and exit from the plastic zone, respectively. Considering now a slab of the strip, of finite thickness, the shape of its contact with the roll may be approximated by

$$y = a + bx \tag{3.57}$$

Then, letting k = constant for a particular slab, equation (3.53) reduces to

$$\frac{dp}{dx} = \frac{\pm p\mu(1+b^2) + 2kb}{(a+bx+C)(1 \mp \mu b)} \tag{3.58}$$

integration of which gives the roll pressure as

$$p = c_1 (a + bx + C)^{\pm \frac{\mu(1+b^2)}{b(1\mp\mu b)}} \mp \frac{2kb}{\mu(1+b^2)} \tag{3.59}$$

where c_1 is a constant of integration.

At the roll-strip interface

$$\sigma_y = p \mp \mu p \frac{dy}{dx} \tag{3.60}$$

Combining equations (3.57) and (3.60) with the Huber-Mises yield criterion then gives another relationship for p and σ_x in the form

$$p = \frac{\sigma_x + 2k}{1 \mp \mu b} \tag{3.61}$$

Eliminating the roll pressure between equations (3.59) and (3.61) gives a relationship for the stress in the direction of rolling as

$$\sigma_x = 2k\left[-1 \mp \frac{b(1\mp\mu b)}{\mu(1+b^2)}\right] + c_1 (1 \mp \mu b)(a + bx + C)^{\pm \frac{\mu(1+b^2)}{b(1\mp\mu b)}} \tag{3.62}$$

Knowing σ_{xi} at the boundary of the ith slab, the constant c_1 may be found. Thus σ_{xi+1} is calculated for the next slab by ensuring horizontal equilibrium between the slabs

$$\sigma_{xi+1} = 2k_i\left[-1 \mp \frac{b_i(1\mp\mu b_i)}{\mu(1+b_i^2)}\right]$$

$$+\left[\sigma_{xi} + 2k_i \pm \frac{2k_i b_i (1\mp\mu b_i)}{\mu(1+b_i^2)}\right]\left(\frac{a_i + b_i x_{i+1} + C}{a_i + b_i x_i + C}\right)^{\pm \frac{\mu(1+b_i^2)}{b_i(1+\mu b_i)}} \tag{3.63}$$

and the roll pressure on slab i is calculated from equation (3.61) as

$$p_i = \frac{\sigma_{xi} + 2k_i}{1 \mp \mu b_i} \tag{3.64}$$

When a straight line segment approximating the roll shape becomes horizontal, i.e. $b = 0$, equation (3.59) is no longer valid. This condition exists near the exit region and may also occur if the roll becomes indented. Thus, equation (3.58) for this case reduces to

$$\frac{dp}{dx} = \pm \frac{\mu p}{a + C} \tag{3.65}$$

and a derivation similar to the one above gives two equations for the roll pressure and the stress in the x direction respectively, as

$$p_i = \sigma_{xi} + 2k_i \tag{3.66}$$

and

$$\sigma_{xi+1} = (\sigma_{xi} + 2k_i)\exp\left[\pm\mu\frac{x_{i+1} - x_i}{a_i + C}\right] - 2k_i \tag{3.67}$$

The second frictional condition given by equation (1.9) is not considered here as it is not believed to be realistic.

Elastic regions. The following derivation is to determine the surface pressure distribution as a function of the gauge profile in the elastic zones. In Fig. 1.1 the elastic entry (compression) and elastic exit (recovery) zones are shown along with the plastic zone. It is assumed that if the vertical pressure loading in the elastic exit zone were removed, the strip would recover elastically to a uniform height h_2 from h_{2e}.

Here the elastic exit is analyzed. The elastic entry is a mirror image i.e. the terms multiplied by the coefficient of friction should be multiplied by -1. The derivation now makes use of the equation of equilibrium in the y direction

$$\sigma_y = p\left(\mu\frac{dy}{dx} + 1\right) \tag{3.68}$$

In the x direction

$$\frac{d\sigma_x}{dx} = -\frac{\sigma_x \frac{dy}{dx}}{y+C} + \frac{p\left(\mu - \frac{dy}{dx}\right)}{y+C} \tag{3.69}$$

Hooke's law for the plane strain case is now used to express the strain in the y direction, integration of which defines the change of thickness, $\delta = h - h_u$, as

$$\delta = \int_0^h \epsilon_y dy = \frac{1-\nu^2}{E}\sigma_y h - \frac{\nu(1+\nu)}{E}\sigma_x h \tag{3.70}$$

Here use was made of the fact that the only independent variable is x. The term h_u designates the undeformed strip thickness. It is equal to h_1 at entry and h_{2e} at exit. In equation (3.70) E is Young's modulus and ν represents the Poisson's ratio of the strip. At the end of the region $h = h_2$ and since no more contact with the roll exists

$$\sigma_y = 0 \tag{3.71}$$

As well, the stress in the x direction must equal the back tension, so at that location

$$\sigma_x h_2 = -T_2 \tag{3.72}$$

where T_2 designates the back tension stress per unit width. Since the total change of thickness at the recovery region is

$$\delta = h_2 - h_u \tag{3.73}$$

where h_2 is the thickness after unloading, the exit thickness becomes

$$h_2 = h_u + \frac{\nu(1+\nu)}{E}T_2 \tag{3.74}$$

Combining equations (3.51) and (3.70) with (3.74) yields

$$\sigma_y = \frac{E[2(y+C)-h_u]+2\sigma_x \nu(1+\nu)(y+C)}{2(1-\nu^2)(y+C)} \tag{3.75}$$

Substituting (3.75) in (3.68) and eliminating the roll pressure between that result and equation (3.69) gives

$$\frac{d\sigma_x}{dx} = -\sigma_x \left[\frac{\frac{dy}{dx}}{y+C} - \frac{\nu\left(\mu - \frac{dy}{dx}\right)}{(1-\nu)(y+C)\left(\mu\frac{dy}{dx}+1\right)} \right]$$

$$+\frac{1}{2}E\frac{[2(y+C)-h_u]\left(\mu-\frac{dy}{dx}\right)}{(1-\nu^2)(y+C)^2\left(\mu\frac{dy}{dx}+1\right)} \tag{3.76}$$

integration of which, with (3.57) gives the stress in the elastic recovery zone, in the x direction as

$$\sigma_x = -\frac{Eh_u(\mu-b)(a+bx+C)^{-1}}{2(1-\nu^2)(\mu b+1)b(\omega-1)}$$

$$+\frac{E(\mu-b)}{(1-\nu^2)(\mu b+1)b\omega} - c_1(a+bx+C)^{-\omega} \tag{3.77}$$

where

$$\omega = 1 - \frac{\nu(\mu-b)}{b(1-\nu)(\mu b+1)} \tag{3.78}$$

and c_1 is a constant of integration.

Knowing σ_{xi} at the boundary of the ith slab, the constant c_1 can be found. This determines the stress σ_{xi+1} at the other side of that slab, which in turn defines the boundary condition for the $i+1^{th}$ slab. Hence,

$$\sigma_{xi+1} = -\frac{Eh_u(\mu - b_i)(a_i + b_i x_{i+1} + C)^{-1}}{2(1-\nu^2)(\mu b_i + 1)b_i(\omega_i - 1)} + \frac{E(\mu + b_i)}{(1-\nu^2)(\mu b_i + 1)b_i\omega_i}$$

$$+\left[-\sigma_{xi} + \frac{E(\mu - b_i)}{(1-\nu^2)(\mu b_i + 1)b_i\omega_i}\right]\frac{(a_i + b_i x_i + C)^{\omega_i}}{(a_i + b_i x_{i+1} + C)^{\omega_i}}$$

$$-\frac{Eh_u(\mu - b_i)(a_i + b_i x_i + C)^{\omega_i - 1}}{2(1-\nu^2)(\mu b_i + 1)b_i(\omega_i - 1)(a_i + b_i x_{i+1} + C)^{\omega_i}} \tag{3.79}$$

Once σ_{xi} is solved for at the slab boundaries, p_i can be calculated from equations (3.68) and (3.75) as

$$p_i = \frac{E[2(a_i + b_i x_i + C) - h_u]}{2(1-\nu^2)(a_i + b_i x_i + C)(\mu b_i + 1)} + \frac{\nu}{(1-\nu)(\mu b_i + 1)}\sigma_{xi} \tag{3.80}$$

For the entry region the expressions for p and σ_x remain the same except for the following changes

$$\mu = -\mu \quad ; \qquad h_2 = h_1 \quad ; \qquad T_2 = T_1$$

The interface between the elastic and plastic zones is determined by checking where and when the yield criterion of Huber-Mises is satisfied - see equations (3.55) and (3.56).

3.2.2 Roll deformation

The schematic loading diagram of a work roll is shown in Fig.3.2. The loading consists of the roll pressure, designated by $p(\phi)$ and the interfacial friction stress, $q(\phi)$. These loads must be balanced by a statically equivalent force system, the location of which would depend on the type of mill considered. For four or six high mills the backup rolls would supply the necessary balance while for two high mills the reaction forces would be applied through the bearings to the roll journals. As the present analysis is concerned with a two dimensional treatment of the work rolls of a two-high mill and roll bending is neglected, the balancing loads are taken to be distributed over the rolls with $\xi = \pi/2$, where ξ represents an angle which determines part of the arc subjected to the balancing loads.

This condition ensures that the midplane of the roll remains undeformed and stationary. Following Michell (1900), the stress and strain distributions in the cylinder in a state of plane strain may be calculated from a stress function, given by

$$\Phi = c_0 r^2 + d_1 r^3 \sin\phi + d_2 r^3 \cos\phi + \sum_{n=2}^{\infty} (a_{1n} r^n + b_{1n} r^{n+2}) \sin(n\phi)$$

$$+ (a_{2n} r^n + b_{2n} r^{n+2}) \cos(n\phi) \tag{3.81}$$

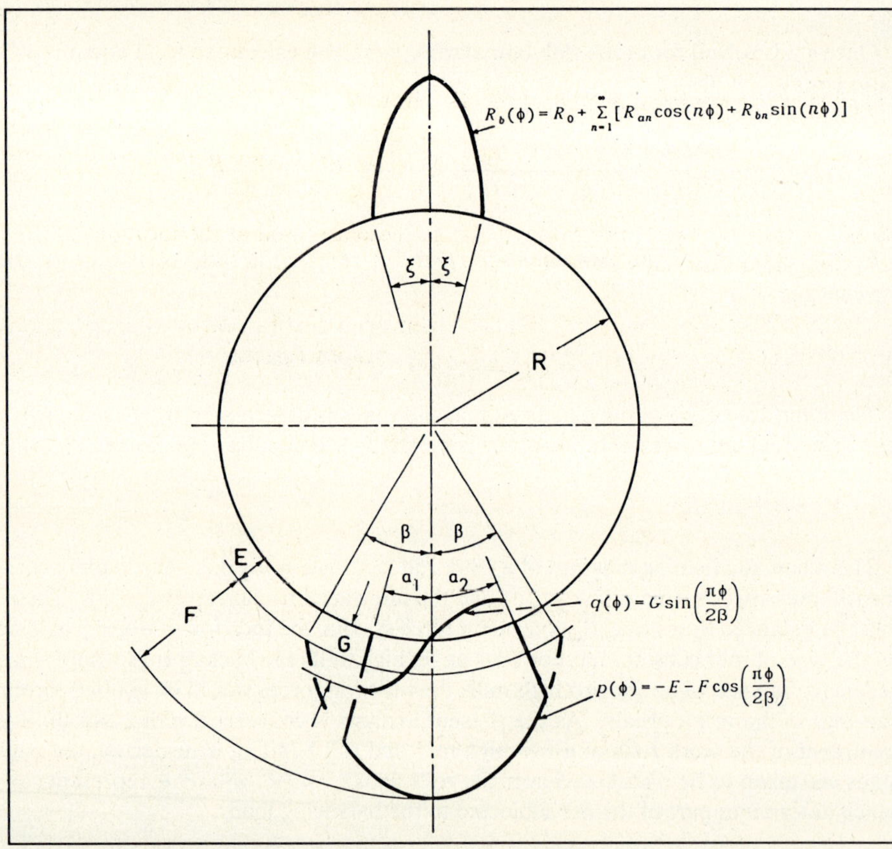

Fig.3.2 Loading diagram of a work roll.

where the coefficients of terms singular at the origin were taken to equal zero. In equation (3.81) r and ϕ are the coordinates in a cylindrical system. The radial and shear stresses are then obtained from

$$\sigma_r = \frac{1}{r}\frac{\partial \Phi}{\partial r} + \frac{1}{r^2}\frac{\partial^2 \Phi}{\partial \phi^2} \tag{3.82}$$

and

$$\tau_{r\phi} = -\frac{\partial}{\partial r}\left(\frac{1}{r}\frac{\partial \Phi}{\partial \phi}\right) \tag{3.83}$$

respectively, subject to boundary conditions, given by

$$\sigma_r = p(\phi) \qquad \text{and} \qquad \tau_{r\phi} = q(\phi) \qquad \text{at} \qquad r = R \tag{3.84}$$

for the contact region. Expressions similar to these are used at the locations of the forces providing the reactions required for balance.

The coefficients in equation (3.81) are determined next by representing the normal and shear loading on the roll - $p(\phi)$ and $q(\phi)$ - in terms of Fourier series

$$p(\phi) = p_{a0} + \sum_{n=1}^{\infty} [p_{an}\cos(n\phi) + p_{bn}\sin(n\phi)] \tag{3.85}$$

$$q(\phi) = q_{a0} + \sum_{n=1}^{\infty} [q_{an}\cos(n\phi) + q_{bn}\sin(n\phi)] \tag{3.86}$$

where the coefficients are obtained from the Euler formulae. For the normal pressure distribution then

$$p_{a0} = \frac{1}{2\pi}\int_{\alpha_1}^{\alpha_2} p(\phi)d\phi + \int_{-\frac{\pi}{2}}^{\frac{\pi}{2}} R_b(\phi)d\phi \tag{3.87}$$

$$p_{an} = \frac{1}{\pi}\int_{\alpha_1}^{\alpha_2} p(\phi)\cos(n\phi)d\phi + \int_{-\frac{\pi}{2}}^{\frac{\pi}{2}} R_b(\phi)\cos(n\phi)d\phi \tag{3.88}$$

and

$$p_{bn} = \frac{1}{\pi}\int_{\alpha_1}^{\alpha_2} p(\phi)\sin(n\phi)d\phi + \int_{-\frac{\pi}{2}}^{\frac{\pi}{2}} R_b(\phi)\sin(n\phi)d\phi \tag{3.89}$$

and for the shear stress distribution

$$q_{a0} = \frac{1}{2\pi}\int_{\alpha_1}^{\alpha_2} q(\phi)d\phi \tag{3.90}$$

$$q_{an} = \frac{1}{\pi}\int_{\alpha_1}^{\alpha_2} q(\phi)\cos(n\phi)d\phi \tag{3.91}$$

and

$$q_{bn} = \frac{1}{\pi}\int_{\alpha_1}^{\alpha_2} q(\phi)\sin(n\phi)d\phi \tag{3.92}$$

In equations (3.87), (3.88) and (3.89) $p(\phi)$designates the roll pressure distribution in the roll gap. It may be expressed as a cosine function in the form

$$p(\phi) = -E - F\cos\frac{\pi\phi}{2\beta} \tag{3.93}$$

shown in Fig.3.2, or alternatively, it may be given as a set of calculated points. In equation (3.93) E and F are constants. The function R_b in equations (3.87), (3.88) and (3.89) represents the reactions required to keep the work roll in equilibrium. This can be calculated by expressing the reactions in terms of Fourier series

$$R_b(\phi) = R_0 + \sum_{n=1}^{\infty} [R_{an}\cos(n\phi) + R_{bn}\sin(n\phi)] \tag{3.94}$$

where again, the coefficients are obtained by the Euler formulae

$$R_0 = -\frac{1}{\pi^2} R_m \xi$$

$$R_{an} = -2\xi R_m \frac{\cos[(-\pi + \psi + \xi)n] + \cos[(-\pi + \psi - \xi)n]}{\pi^2 - 4\xi^2 n^2}$$

$$R_{bn} = -2\xi R_m \frac{\sin[(-\pi + \psi + \xi)n] + \sin[(-\pi + \psi - \xi)n]}{\pi^2 - 4\xi^2 n^2}$$

In the above equations ψ is the angle between the resultant reaction force and the vertical axis, R_m is the amplitude of the reaction force and ξ represents half of the angle over which reaction R_b is distributed. It is noted that for a 4 high mill the value of ξ may be determined from the Hertz contact stresses. However, as was mentioned above, for a two high mill the reactions are represented by letting $\xi = \pi/2$ which in fact indicates distribution of those forces across a diametral plane of the roll.

In equations (3.90),(3.91) and (3.92) the shear stress distribution is represented by

$$q_b(\phi) = G \sin\frac{\pi\phi}{2\beta} \tag{3.95}$$

The values E, F and G are indicated in Fig.3.2. As shown, the maximum value of the roll pressure is given by $E + F$; the period is 2β; entry and exit locations are designated by α_1 and α_2 respectively. G stands for the maximum value of the friction stress. Finally, the coefficients of the stress function expansion - and hence, those for the stresses and strains - are obtained as

$$c_0 = \frac{p_{a0}}{2}$$

$$d_1 = \frac{p_{b1}}{2r_b} = -\frac{q_{a1}}{2r_b}$$

$$d_2 = \frac{p_{a1}}{2r_b} = \frac{q_{b1}}{2r_b}$$

$$a_{1n} = \frac{np_{bn} - (n-2)q_{an}}{2n(n-1)r_b^{n-2}} \tag{3.96}$$

$$b_{1n} = \frac{p_{bn} - q_{an}}{2(n+1)r_b^n}$$

$$a_{2n} = -\frac{np_{an} - (n-2)q_{bn}}{2n(n-1)r_b^{n-2}}$$

$$b_{2n} = \frac{p_{an} + q_{an}}{2(n+1)r_b^n}$$

and the plane strain form of Hooke's law is used to determine the strain components.

3.2.3 Roll Force and Roll Torque

The roll separating force is obtained by integrating the roll pressure distribution over the arc of contact which includes the elastic compression and recovery zones. The expression

$$F = p[\int_{x_1}^{x_n}\left(1 + \mu\frac{dy}{dx}\right)dx + \int_{x_n}^{x_2}\left(1 - \mu\frac{dy}{dx}\right)dx$$
$$+ \int_{x_1}^{x_n}\left(\mu - \frac{dy}{dx}\right)dx + \int_{x_n}^{x_2}\left(\mu + \frac{dy}{dx}\right)dx] \tag{3.97}$$

is used where x_1 and x_2 define the boundaries of the roll gap. The roll torque is computed by

$$M = p\int_{x_1}^{x_n}\left[x - y\frac{dy}{dx} + \mu\left(y + xy\frac{dy}{dx}\right)\right]dx$$
$$-p\int_{x_n}^{x_2}\left[x - y\frac{dy}{dx} - \mu\left(y + x\frac{dy}{dx}\right)\right]dx \tag{3.98}$$

When a constant coefficient of friction is used, the location of the neutral plane is determined in the usual way, that is, by integrating the relevant equations from the exit and entry and searching for the place where pressures for both forward and backward slip zones are equal. If $\mu = \mu(x)$, obtained from experiments, is used the location of zero interfacial shear is supplied with the information. The calculations then begin at entry and the validity of the solution is proven by satisfying the prescribed boundary condition at exit without actually enforcing it.

3.3 Comparison of Mathematical Models of Flat Rolling

3.3.1 Comparison of Assumptions in Various Models

Six mathematical models are reviewed in this section in terms of the assumptions and simplifications made either during their development and derivation, or when roll pressures, forces and torques are calculated. The models are due to Orowan (1943), Bland and Ford (1948), Sims (1954), Ford and Alexander (1963) and Tselikov (1939). The refined model developed by Roychoudhuri and Lenard (1984) and described in Section 3.2 is also included in the comparisons. The components of any model, as pointed out earlier in Section 3.1, should include a mathematical description of the:

(a) equations of motion of the deforming metal,
(b) equation of equilibrium of the work rolls,
(c) boundary/initial conditions,
(d) material's resistance to deformation.

All six models treat flat rolling as a one-dimensional problem. All assume that the only independent variable is the coordinate in the direction of rolling. All assume the problem to be quasi-static and none includes inertia terms. As well, all assume that homogeneous compression of the strip occurs and that the flow of the strip is occurring in the plane of rolling, that is, plane strain flow is present and there is no spread. The differences become apparent when one examines the way boundary conditions are applied by each model. These boundaries involve the roll-strip interface as well as the entry and the exit regions.

Considering the roll-strip interface, two aspects are to be noted. One is the contour of the interface, the other is friction. At the entry and exit regions, elastic compression and recovery need to be analyzed.

Roll-strip contour. The traditional approach here is to assume that the roll contour remains circular and that the radius of curvature becomes larger than the original roll radius. Hitchcock's (1935) formula is then used to calculate the deformed roll radius. Orowan (1943), Bland and Ford (1948) and Sims (1954) recommend that method of calculation. It was pointed out in some detail by Roychoudhuri and Lenard (1984) however, that Hitchcock's formula does not predict the deformed roll shape very well. To overcome this, in Section 3.2.2 the shape of the roll-strip interface is determined by going back to first principles. The theory of elasticity is used to determine the flattening of an originally circular, solid steel cylinder due to the action of normal and shearing forces over a portion of its cylindrical surface. The resulting formulae then are matched with the analysis of the deforming strip.

A comparison of the ratio of the projected contact length, l_d', to the one based on rigid rolls and rigid-plastic strips, l_d, as a function of reduction, obtained by Hitchcock's (1935) formula, by the complex variable method of Pietrzyk and Zakrzewski (1978) and by the integration of the biharmonic equation (Roychoudhuri and Lenard, 1984) is given for a low carbon and an alloy steel in Fig.3.5. The ratio l_d'/l_d is plotted on the ordinate and the reduction is given along the abscissa. It is observed that the assumptions of elliptical pressure distribution and circular, cylindrical rolls underestimate roll flattening significantly for both steels and that the method of complex variables and the integration of the biharmonic equation give results that are in close agreement.

Interfacial friction. All six techniques of calculation assume that one of the following frictional conditions exists between the roll and the strip:

(a) the coefficient of friction is constant,
(b) sticking friction is present.

Evidence can be found in the technical literature that neither condition is correct - see for example van Rooyen and Backofen (1957), Al-Salehi et al. (1973), Lim and Lenard (1984) and Karagiozis and Lenard (1985). It is the refined model of Section 3.2 that can be most easily modified to include realistic frictional conditions. In that model digitally or functionally supplied forms of the variation of the frictional coefficient may be used.

Elastic entry and exit. Five of the conventional models consider that the theoretical point of entry to the roll gap is at the intersection of the undeformed strip and the roll. Exit is assumed to occur at a line connecting the roll centres. It is, of course, recognized that these assumptions are untrue. Researchers then use the correction formulae, discussed by Alexander (1972), to include the effects of elastic entry and exit on the predicted roll forces and torques. Roychoudhuri and Lenard (1984) again make use of first principles to include the entry and exit

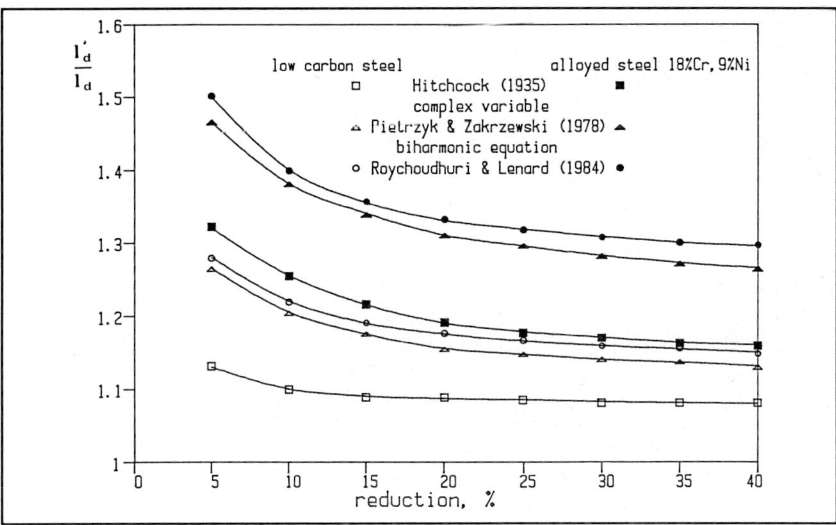

Fig.3.3 Comparison of methods for roll flattening calculations.

regions in the model. Elasticity theory and Hooke's law are employed and the resulting elastic stress distributions are coupled with the plastic regions by using the Huber-Mises criterion of plastic flow.

Simplifying assumptions. The Orowan (1943) model claims to include as few simplifying assumptions as possible to yield an accurate solution. The price one must pay for using that method is the need for a numerical solution. Since all six methods are, by necessity, iterative, their numerical solutions require long computer times. To overcome the need for a numerical solution, Sims (1954) introduced simplifications, all based on the fact that roll gap angles are small compared to unity, given by equations (3.21) and (3.23). He also assumed that $\tau = k =$ constant in the roll gap. Bland and Ford (1948), in addition to the above, assumed that the roll pressure equals the flow strength. Ford and Alexander (1963) introduced two empirical formulae for the computation of the roll forces and torques. A listing of these assumptions is presented in Table 3.1. Note that Sims used Orowan's correction for flow in inclined planes.

Convergence. In the traditional models of flat rolling, the differential equation of equilibrium is integrated for the roll pressure starting from entry; then, with appropriate changes, from the exit and the neutral point is taken to coincide with the location of the intersection of the

method	roll deformation	entry/exit	friction	small angles	convergence for $R/h \gg 1$
Orowan	Hitchcock	correction	$\mu = \text{const}$	yes	no
Bland & Ford	Hitchcock	correction	$\mu = \text{const}$	yes	no
Sims	Hitchcock	correction	$\mu = \text{const}$	yes	no
Ford & Alexander	Hitchcock	correction	$\mu = \text{const}$	yes	no
Tselikov	Hitchcock	correction	$\mu = \text{const}$	yes	no
Refined model	elasticity	elasticity	$\mu = \mu(\phi)$	no	yes
Assumptions common to all:	- homogeneous compression (except Sims) - no inertia forces - one dimensional treatment - plane strain flow				

Table 3.1 Comparison of assumptions made in various models.

two curves. The roll pressure, thus obtained, is referred to as the friction hill. It is known that the friction hill is an unrealistic representation of actual roll pressures. As well, when reduction of hard, thin strips is considered, using large roll diameter-to-strip thickness ratios, the models of Orowan (1943), Bland and Ford (1948), Sims (1954) and Tselikov (1939) fail to converge. In Section 3.2, this problem is solved by using a coefficient of friction in either of two ways. If the actual variation of the coefficient is known, it is supplied to the model, slab by slab. If that is not available, a smooth changeover at the no-slip point is assumed. Both methods cause the refined model described in Section 3.2 to yield converging solutions when all other models fail.

3.3.2 Statistical Analysis of Predictive Capabilities (Murthy and Lenard, 1982)

In this section two techniques of computation and two material flow strength formulations are tested for their ability to predict consistently and accurately separating forces and mill powers in hot strip rolling. These are referred to as methods 1 to 4 and are listed below:

(a) Orowan's (1943) model for the mechanics of rolling, together with a modified version of Shida's (1974) flow strength formulation, described by equation (2.2),

(b) Orowan's model with Ekelund's flow strength formulation, the latter being presented by Wusatowski (1969),

(c) Ford and Alexander's (1963) model for the mechanics together with the Shida's flow strength formulation,
(d) Ford and Alexander's model with Ekelund's flow strength formulation.

Materials. Low carbon steel and HSLA steel were used in the laboratory experiments. Their chemical compositions are given in Table 3.2.

	C	Mn	S	P	Si	Cr	Ni	Cu	Mo	Nb
LC	0.19	0.82	-	-	-	0.1	-	-	-	-
HSLA	0.18	1.14	0.028	0.04	0.02	0.04	0.01	0.02	0.01	0.022

Table 3.2 Chemical composition of steels used in the laboratory experiment.

Statistical evaluation. The results of the calculations are given in Tables 3.3,3.4 and 3.5. Tables 3.3 and 3.4 involve data taken off industrial logs while Table 3.5 is concerned with data generated on a small, experimental rolling mill. In each case, the numbers given represent the percentage difference of measured and predicted values. Table 3.3 shows the means and the standard deviations of the errors involving separating force calculations. Results for the first four stands of the finishing mill are given. Low carbon steel and HSLA steel are contrasted by having respective numbers placed in corresponding rows. Table 3.4 repeats the above but is concerned with the mill power. Finally, Table 3.5 gives similar results of computations for the laboratory data.

Some general observations may be made and these are listed below:

a. In all of the analyses of industrial data, the variation of the percent error with stand number shows a definite and identical trend. The error becomes more positive in the first through third stand and dips (i.e. becomes less positive) in the fourth stand. Note that the percentage error for both force and power is calculated as

$$\bar{\delta} = \frac{\text{measured value} - \text{predicted value}}{\text{predicted value}} 100\%$$

from which it is evident that an increasing percentage error implies a predicted value which is decreasing with respect to the measured value. It is well known that the strain rate increases with stand number while the average temperature decreases. In fact both these factors result

pass no.	steel	method 1		method 2		method 3		method 4	
		$\bar{\sigma}$, %	$\bar{\delta}$, %	$\bar{\sigma}$, %	$\bar{\delta}$, %	$\bar{\sigma}$, %	$\bar{\delta}$, %	$\bar{\sigma}$, %	$\bar{\delta}$, %
1	LC	15.90	9.68	4.61	12.95	-49.19	17.29	-51.30	20.07
	HSLA	19.97	12.04	-8.49	21.96	-41.39	19.17	-63.87	21.28
2	LC	26.65	6.74	13.67	9.18	-26.81	11.62	-32.97	13.72
	HSLA	25.73	9.27	-1.52	14.18	-27.73	14.01	-50.60	20.57
3	LC	45.03	6.32	34.93	8.40	7.42	10.63	1.53	15.12
	HSLA	45.22	10.36	20.64	13.69	4.08	12.39	-12.58	18.48
4	LC	41.49	0.08	30.71	11.39	4.07	15.70	-2.18	18.25
	HSLA	36.55	13.21	14.40	16.50	-3.37	18.80	-20.52	24.61

Table 3.3 Average error and standard deviation for roll force calculations in an industrial strip mill.

pass no.	steel	method 1		method 2		method 3		method 4	
		$\bar{\sigma}$, %	$\bar{\delta}$, %	$\bar{\sigma}$, %	$\bar{\delta}$, %	$\bar{\sigma}$, %	$\bar{\delta}$, %	$\bar{\sigma}$, %	$\bar{\delta}$, %
1	LC	2.83	12.39	-10.27	14.51	-35.80	25.04	-36.47	26.59
	HSLA	5.72	20.92	-28.99	25.94	-31.82	24.60	-51.99	25.06
2	LC	20.41	8.70	5.74	11.39	-6.09	17.15	-10.18	18.60
	HSLA	17.64	11.78	-12.86	19.85	-6.48	12.29	-25.34	15.92
3	LC	25.89	10.60	11.83	13.44	8.48	16.40	4.21	17.82
	HSLA	26.97	21.02	-6.33	20.07	9.34	13.12	-6.29	17.94
4	LC	24.94	10.79	10.53	13.75	11.07	16.62	8.85	18.98
	HSLA	17.50	23.16	-12.14	24.71	7.18	16.77	-8.17	21.55

Table 3.4 Average error and standard deviation for mill power calculations in an industrial strip mill.

in an increasing value of material flow strength. Of the four computational schemes used for predicting forces and powers, the two using Ekelund's material formulation do not acknowledge any influence of the strain rate. This means that although the material is actually becoming harder, the predictive techniques are not keeping track of those increases resulting

	steel	method 1		method 2		method 3		method 4	
		$\bar{\sigma}$, %	$\bar{\delta}$, %	$\bar{\sigma}$, %	$\bar{\delta}$, %	$\bar{\sigma}$, %	$\bar{\delta}$, %	$\bar{\sigma}$, %	$\bar{\delta}$, %
force	LC	9.95	17.21	-43.56	12.49	-38.83	21.53	-101.76	23.71
	HSLA	-34.86	26.86	-57.25	17.65	-129.90	47.93	-128.10	35.45
power	LC	15.02	18.73	-19.55	30.14	12.50	20.20	-28.10	35.63
	HSLA	6.08	14.50	-6.09	32.32	-17.14	20.27	-26.58	64.36

Table 3.5 Average error and standard deviation for roll force and mill power calculations in laboratory conditions.

in a predicted value which is decreasing with respect to the measured one. Although the other techniques using Shida's formulation as the basis for material behaviour do acknowledge a variation of predicted forces and powers with increasing strain rates, it could be argued that it is possible they do not do so sufficiently well, resulting in predicted values that are less than the actual ones. A similar argument may hold for the effect of decreasing temperatures. Even though all four models contain a temperature factor it is still possible that the effect of temperature is not represented sufficiently well, causing the predicted forces and powers to lag behind the measured values.

In all calculations it has been assumed that a condition of sticking friction exists. A study of recent, relevant literature reveals that in most cases of hot rolling this is a questionable assumption. At the lower temperatures encountered at the fourth stand it is likely that this assumption is certainly no longer valid and instead the friction stress may be governed by a relation of the form

$$\tau = p\mu(p, T, v_s, \text{material}, \text{surface condition})$$

where τ is the frictional stress, p is the normal pressure, μ is the coefficient of friction as a function of normal pressure, temperature T, velocity of sliding v_s and the contacting materials. Assuming that this is the case it follows that at the fourth stand the numerical technique which continues to use the sticking friction condition will predict forces and power that have increased with respect to the measured values. This would explain the reversed trend occurring at the fourth stand.

b. On studying the variation of the standard deviation with mill stand number it is observed that the standard deviation is least in the second and third stands and increases in the first and fourth stands. The magnitude of the effect varies but the general trend is always the same. It is impossible to pinpoint any one reason for this effect. It could be caused by a combination of one or more of the following possibilities. It seems reasonable to believe that the first stand which receives the strip directly from the roughing mill would receive a greater variation in the strip thickness than the subsequent stands. Since the data logging device used in the industry probably reads an instantaneous value of strip thickness, a variation in the input thickness would result in a variation in the percentage error calculated for the first stand. The difference between the inlet side looper tension and the outlet side looper tension at any stand represents the net pull on the strip. This net pull is always the largest in the first stand since the inlet side tension is always zero. The larger net pull will result in a shift of the neutral point closer to one end of the arc of contact in the roll gap. This shift in the neutral point may make the calculations for that stand more susceptible to variations in frictional stresses. As the friction effects may vary randomly between sticking friction and Coulomb friction, due to the presence of scales in the roll gap, the error in the predictions will vary too. As has been observed earlier, at the fourth stand it is likely that the nature of friction has changed. This could help explain the greater variance in the predictions at the fourth stand. The second, third and fourth finishing mill stands operate in a temperature zone which is in the vicinity of the border between hot and cold rolling. The exact point of occurrence of this transition is not clear. However, it is possible that this transition occurs close to the temperatures normally encountered at the fourth stand, i.e. conditions at the fourth stand may alternate between hot rolling and cold rolling. Further, the standard deviation inherent in the industry data is unknown. This may have, to a greater or lesser extent, been reflected in the results.

c. The mean error between the predicted value and the observed value represents the accuracy of the prediction whereas the standard deviation in this error represents the consistency of the prediction. While the accuracy determines how close the predicted value comes to the measured value, the precision determines the reproducibility of the results. In judging the usefulness of a predictive technique, both these factors should be taken into consideration. However, it should be borne in mind that an inaccurate result can be corrected for by an additional multiplying factor, but the effect of an inconsistent result can never be corrected. Hence, in any predictive technique, good accuracy is desirable and good consistency is essential.

d. It should be pointed out here that normally one would expect the measured values to be greater than the predicted power values because the latter do not take into consideration the efficiency of the bearings and the power transmission train. Adding a factor to the predicted values to compensate for this effect would make the errors in Ford and Alexander's predictions higher, that is, more negative. It would however help to bring the values predicted by the Orowan method closer to the actual measured values.

e. It is noted that for the laboratory data, the distinction between the Ford and Alexander technique and the Orowan formulation becomes ever more pronounced, with errors in force predictions reaching as high as 100% and more. The reason for this phenomenon is that the Ford and Alexander technique does not consider the effect of looper tension explicitly while the Orowan technique does. In the laboratory testing, no back or front tensions have been applied to the strip. Presumably then, the Orowan technique can adjust to a value of zero looper tension, whereas the Ford and Alexander method cannot.

f. On the whole, for industrial data as well as for laboratory data, standard deviations for the HSLA steel are higher than corresponding values for the mild steel. This is probably due to the fact that the material flow strength models used do not consider the strengthening effects of microalloying elements in the HSLA steel, thus introducing an extra element of variability that is not allowed for in the predictive models.

g. Contrary to what may be expected, mean errors for HSLA steel are more negative than the corresponding values for mild steel. Since the material models used do not consider the the strengthening effect of the microalloying elements, one would expect the predicted forces and powers to decrease with respect to the measured values, thus resulting in the error becoming more positive. This does not appear to be the case.

h. In general, it may be noted from the results both for industrial and laboratory data that the accuracy and precision seem to go hand in hand with the relative complexity of the predictive scheme. Thus the Orowan and Shida technique appears to be the most consistently accurate, followed by that of Orowan and Ekelund, then by Ford and Alexander and Shida and, finally, Ford and Alexander and Ekelund.

3.4 Further Substantiation of the Predictive Capabilities of One-Dimensional Models

Convergence. One of the often repeated comments regarding Orowan's (1943) method of analysis is that for large roll radius-to-strip thickness ratios it fails to converge. Calculations were performed to substantiate those comments. Also, the more refined technique of Roychoudhuri and Lenard (1984) was tested for its speed of convergence. Taking a metal whose true stress - true strain curve is described by

$$\sigma = 593.5(1 + 78.3\epsilon)^{0.067}$$

an entry thickness of 3.15 mm, roll radius of 125 mm and a coefficient of friction of 0.25, Orowan's (1943) model as computerized by Alexander (1972), failed to converge when reductions in excess of 50% were attempted. The model presented in Section 3.2, using $\mu = \mu(\phi)$, converged to realistic values in nine iterations.

Strain distribution in the work rolls. In this section the roll deformation calculations presented by Roychoudhuri and Lenard (1984) are compared to measured results. In order to monitor the variation of radial strain during rolling, two strain gauges were cemented on the face of the top roll. One of the gauges was located at a radial distance of 25.8 mm from the centre, the other somewhat closer, at 24.6 mm. The signals from the gauges were recorded on a two channel Watanabe strip chart recorder. A strip of 1100 - T0 aluminium alloy of 50 mm width was rolled to a reduction of 5%, the roll separating force being measured as 17700 N. The experimental results and analytical predictions of Roychoudhuri and Lenard (1984) are given in Figs 3.4 and 3.5. In both figures the radial strains are plotted on the ordinates while the

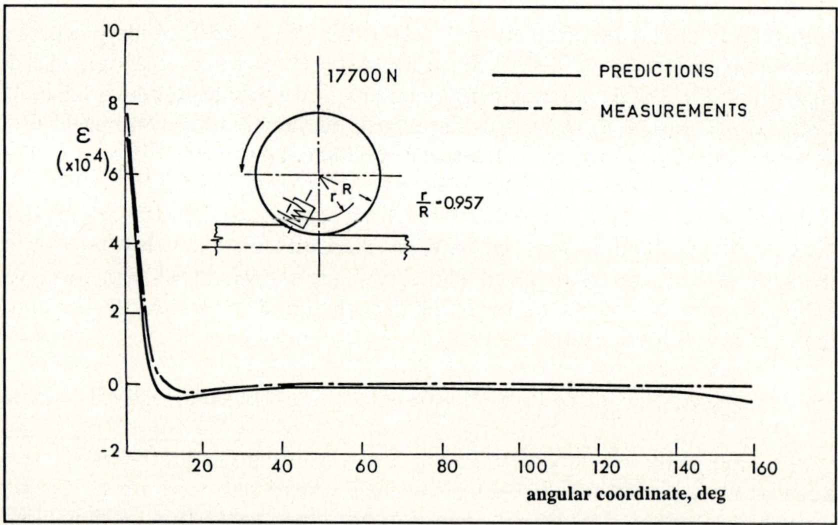

Fig.3.4 Comparison of measured and calculated strains in the roll.

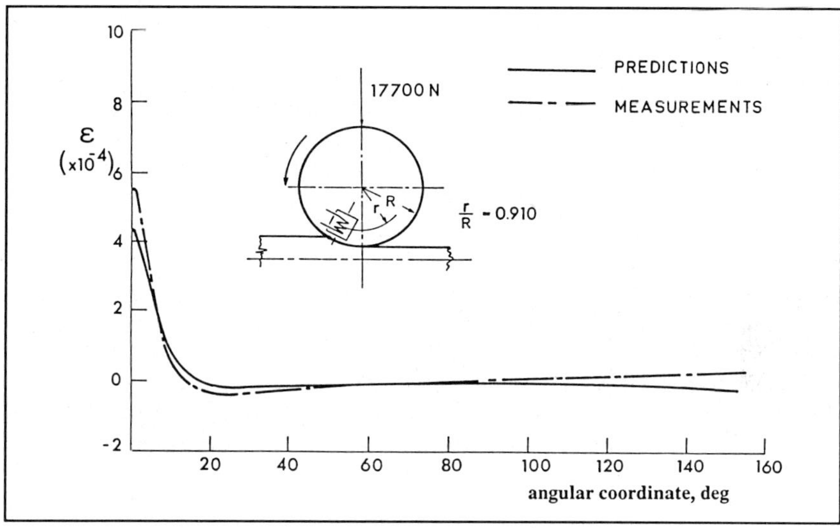

Fig.3.5 Comparison of measured and calculated strains in the roll.

angular variable is given on the abscissa. The origin locates a strain gauge passing through the position of maximum roll pressure. In Fig.3.4, the strain gauge was placed as close to the edge as possible. It may be observed from the figures that the analytical predictions for the radial strain distribution are quite good. The largest difference, 21%, is found to occur at the origin for the gauge located closer to the roll centre. The predictions and measurements agree very closely for the gauge near the roll surface. It may be argued that, since roll bending is neglected in the two dimensional analysis, the difference between predictions and measurements would grow as one approaches the roll centre line. At locations away from the position of maximum strains both gauges indicate that the analysis is reliable. The differences are never more than 15 microstrains.

Cold rolling. In this section the predictive capabilities of the mathematical models, described earlier, are compared to the experimental data of Shida and Awazuhara (1973) obtained during cold rolling of two types of steel strips with the stress-strain relationships given by

steel 1: $\sigma = 324(1 + 19.2\epsilon)^{0.295}$

steel 2: $\sigma = 358(1 + 22.7\epsilon)^{0.3}$

The roll radius was 65 mm. The frictional coefficient used in all calculations was 0.08. The entry thickness of the strip was 1 mm for steel 1 and 0.5 mm for steel 2. The results of measurements and calculations are given in Figs 3.6 - 3.9. In all of the figures roll forces and torques per unit width are plotted on the ordinates and the reduction in percent is given along the abscissa. A general conclusion is that, at least at low reductions, all five techniques of computations give results that are "in the ballpark". Orowan's (1943) technique as programmed by Alexander (1972) proved to be reasonable in predicting both forces and torques. The method of Roychoudhuri and Lenard (1984) did not distinguish itself, as its torque predictions for both steels were poor. The Bland and Ford (1948) analysis gave results very close to those of Orowan (1943).

The coefficient of friction was taken to equal 0.08 in all of these calculations. As is often the practice, this coefficient could be treated as a free parameter, and the predictions of the models could be improved by a judicious choice of μ. The frictional coefficient may also be measured however, and when the experimentally established values are used in the calculations, the predictive capabilities of the models increase significantly. This is shown in Table 3.6 where the data regarding μ, developed by Lim and Lenard (1984) and Karagiozis and Lenard (1985), (see also Chapter 2), is used to demonstrate the validity of that claim. When the coefficient of friction values are digitized and included in the model, noticeable improvements in the predictive capabilities are observed.

It may be concluded at this stage that, as far as the models' ability to compute roll separating forces and roll torques is concerned, all perform in a satisfactory manner. Variations are observed; however, none is considered to be significant. Inclusion of experimentally determined data concerning friction at the roll-strip interface enhances the ability of the model of Roychoudhuri and Lenard (1984) to calculate roll forces. It is probably premature to attempt to pick a best model. A more likely approach may be to choose the most appropriate model for a particular situation. For large roll radius - to - strip thickness ratios the slab models appear to be satisfactory. For large reductions roll deformation, realistic frictional conditions and the effects of the elastic entry/exit regions need to be considered.

Hot rolling. Attempts to provide answers to three questions are described in this section. The first is, how well could the model developed by Roychoudhuri and Lenard (1984) and tested against laboratory data (Lenard, 1986b), predict forces and torques measured in an industrial situation? Second, to what degree would the results depend on the manner of representation of constitutive data? Finally, how would the traditional method of Orowan (1943) fare in comparison to the more refined one, presented in Section 3.2? Several steps are involved

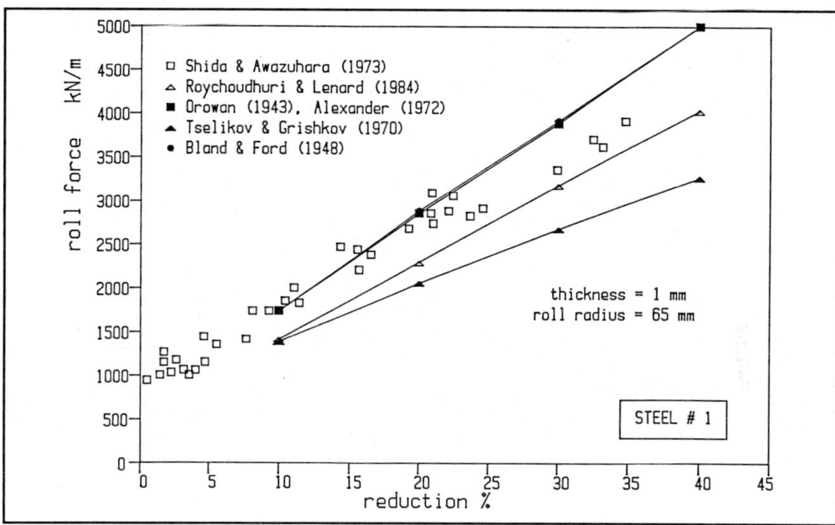

Fig.3.6 Measured and calculated roll force for steel #1.

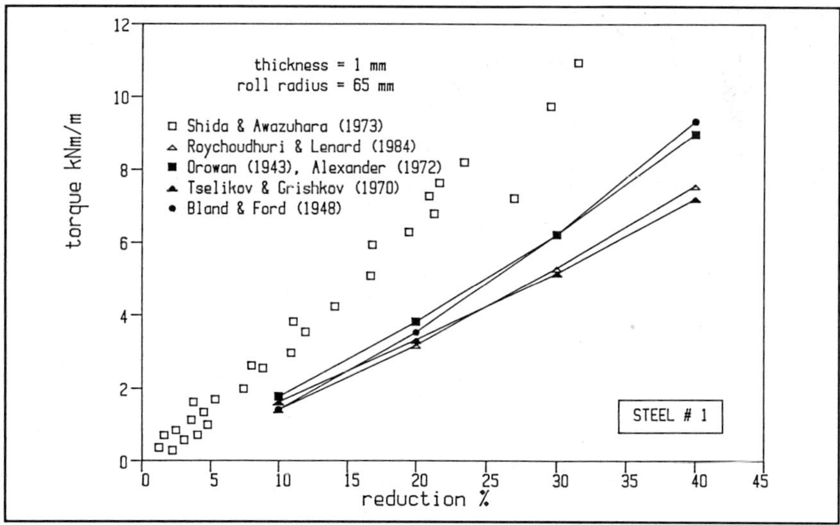

Fig.3.7 Measured and calculated roll torque for steel #1.

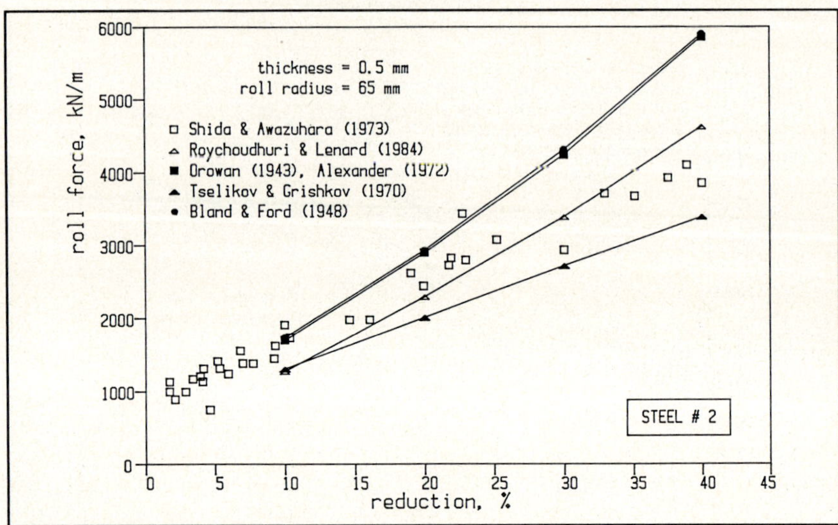

Fig.3.8 Measured and calculated roll force for steel #2.

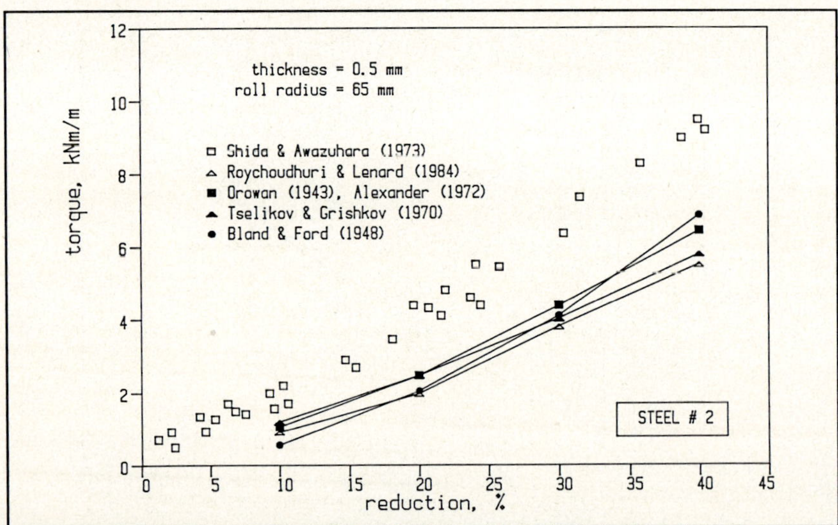

Fig.3.9 Measured and calculated roll torque for steel #2.

material	measured	Orowan (1943) $\mu = const$	refined model $\mu = const$	refined model $\mu = \mu(\phi)$
Al 5052 - H34	3651	4956	5098	4061
Al 6061 - T6	3869	4752	4926	3845
Al 2024 - T3	4460	4060	4364	4323

Table 3.6 Comparison of measured and calculated roll forces, N/mm.

in the attempt to provide answers to these questions. These include obtaining reliable data from production runs, conducting compression tests at constant strain rates to determine the resistance of the rolled material to deformation and combining that data with models of flat rolling to calculate the rolling variables. During the computations decisions regarding the manner of representation of constitutive data are required. In the analysis of results the dependence of the predictive capabilities of the two models of flat rolling on those decisions are also evaluated. Three techniques to model material behaviour are employed. These are the development of a multidimensional databank, use of average values and derivation of empirical equations. Each of these is then combined with either Orowan's (1943) or Roychoudhuri and Lenard's (1984) model of flat rolling and the computed roll powers and forces are compared to production data provided by Sidbec-Dosco Co. Ltd. of Canada. The data includes all of the parameters and variables necessary for comparison to the predictions of the analyses: temperature, reduction, roll force and mill power, roll diameter (656mm), roll rpm and strip width (965 mm). Table 3.7 shows the pass schedules for a low carbon steel while Table 3.8 presents those for a vanadium bearing steel. The chemical compositions of the steels are given in Table 2.3, Section 2.1. The average strain rates in all passes have been calculated and are given in the last columns of Tables 3.7 and 3.8. No front or back tensions

pass	h_1 mm	h_2 mm	T ^oC	rpm	power kW	force kN/mm	$\dot{\epsilon}$ s^{-1}
1	16.5	10.1	900	87	6699	15.2	32.0
2	10.1	6.7	900	136	6278	13.8	57.1
3	6.7	4.7	900	145	4140	12.2	69.3
4	4.7	3.5	900	85	1656	11.8	43.9
5	3.5	3.1	900	138	1208	5.5	49.2

Table 3.7 Pass schedule for the low carbon steel.

pass	h_1 mm	h_2 mm	T °C	rpm	power kW	force kN/mm	$\dot{\epsilon}$ s^{-1}
1	16.5	10.1	935	94	6732	15.9	34.8
2	10.1	6.7	935	135	6143	13.2	56.4
3	6.7	4.7	927	143	4140	12.2	67.9
4	4.7	3.5	921	146	2726	10.7	73.8
5	3.5	3.1	877	141	1346	6.1	53.0

Table 3.8 Pass schedule for the vanadium steel.

are used in modelling and in all computations the strip is assumed to have recrystallized completely in between passes. The results of compression tests presented in Section 2.1.3 are used to determine the material's yield strength.

Prediction of roll forces and powers. The results of the computations of the roll forces and powers are given in Tables 3.9 and 3.10 for the carbon steel and in Tables 3.11 and 3.12 for the vanadium bearing steel. The measured values are also shown in the tables. The parameters in the constitutive equations, used in the modelling, determined by interpolating or extrapolating from the values of Table 2.4 for the mean strain rate in each pass are presented in Table 3.13. Also included are the values of the mean flow strength for each run, as obtained by equation (2.10).

pass	force	Orowan (1943)		refined model		
	N/mm^2	$\bar{\sigma}$	$\sigma = \sigma_{po}(1 + B\epsilon)^n - A\epsilon^m$	databank	$\bar{\sigma}$	$\sigma = \sigma_{po}(1 + B\epsilon)^n - A\epsilon^m$
1	15.2	16.9	17.0	16.9	14.6	20.1
2	13.8	13.7	14.0	13.5	11.8	12.6
3	12.1	11.8	12.2	11.4	10.0	10.9
4	11.8	9.7	10.1	9.3	8.2	9.0
5	5.5	4.0	4.1	3.6	3.4	3.8

Table 3.9 Prediction of roll forces during rolling of the low carbon steel.

The coefficient of friction is assumed to be 0.2. This appears to be quite low, especially when the recommendation of Alexander (1972) - to assume that sticking friction exists in hot rolling - is considered or Ekelund's or Bakhtinov's formulae, as presented by Wusatowski (1969), are used. Previous studies, however, showed that a value of $\mu = 0.3$ gave the best results when modelling hot rolling of Nb steel was done (Lenard, 1986b) and this was reinforced in the present study where the use of $\mu = 0.2$ led to the least amount of difference between predictions and measurements. The results of Sparling (1977) and Roberts (1974) also point to this finding.

pass	power	Orowan (1943)		refined model		
	kW	$\bar{\sigma}$	$\sigma = \sigma_{po}(1 + B\epsilon)^n - A\epsilon^m$	databank	$\bar{\sigma}$	$\sigma = \sigma_{po}(1 + B\epsilon)^n - A\epsilon^m$
1	6699	6155	6058	5966	5328	6885
2	6278	5607	5544	5049	4729	4958
3	4140	3910	3821	3619	3337	3286
4	1656	1431	1385	1365	1219	1162
5	1208	543	532	463	481	479

Table 3.10 Prediction of mill powers during rolling of the low carbon steel.

pass	force	Orowan (1943)		refined model		
	N/mm^2	$\bar{\sigma}$	$\sigma = \sigma_{po}(1 + B\epsilon)^n - A\epsilon^m$	databank	$\bar{\sigma}$	$\sigma = \sigma_{po}(1 + B\epsilon)^n - A\epsilon^m$
1	15.9	16.6	14.2	16.6	14.4	17.6
2	13.2	13.3	13.8	13.3	11.4	14.2
3	12.2	11.1	11.5	11.4	9.5	11.9
4	10.7	8.8	9.1	9.2	7.4	9.5
5	6.1	4.3	4.5	3.9	3.6	4.6

Table 3.11 Prediction of roll forces during rolling of the vanadium bearing steel.

pass	power	Orowan (1943)		refined model		
	kW	$\bar{\sigma}$	$\sigma = \sigma_{po}(1 + B\epsilon)^n - A\epsilon^m$	databank	$\bar{\sigma}$	$\sigma = \sigma_{po}(1 + B\epsilon)^n - A\epsilon^m$
1	6732	6581	7226	6156	5636	6495
2	6142	5378	5231	5241	4533	5330
3	4140	3638	3554	3525	3241	3526
4	2726	2230	2214	2052	1828	2056
5	1346	640	624	540	528	558

Table 3.12 Prediction of mill powers during rolling of the vanadium bearing steel.

material	$\dot{\epsilon}$ s^{-2}	σ_{p0} N/mm^2	B	n	A N/mm^2	m	$\bar{\sigma}$ N/mm^2
low carbon steel	32.0	101	55.81	0.5018	560.90	0.9078	223.21
	57.1	101	62.10	0.4967	562.15	0.8540	220.05
	69.3	101	64.66	0.4942	563.02	0.8411	216.05
	43.9	101	58.98	0.4994	561.39	0.8773	208.58
	49.2	101	60.37	0.4983	561.60	0.8639	174.31
vanadium steel	34.8	100.5	53.97	0.4698	436.90	0.8944	219.02
	56.4	101	67.97	0.4605	453.83	0.7869	215.90
	67.9	101	75.38	0.4556	462.78	0.7300	207.39
	73.8	101	79.19	0.4531	467.37	0.7008	196.30
	53.0	101	65.75	0.4620	451.15	0.8040	174.63

Table 3.13 Coefficients in equation (2.9).

CHAPTER 4

THERMAL-MECHANICAL FINITE-ELEMENT MODELLING OF FLAT ROLLING

The conventional analytical methods discussed in the previous chapters accounted for mechanical events in the deformation zone. Thermal phenomena were considered only by introducing the temperature-dependent properties of the rolled metal. Realistic analysis of the flat rolling process indicates however, that the events taking place during a pass should be properly looked at as a thermal-mechanical problem. Arbitrary separation of these components may lead to erroneous appreciation of the variables and parameters that characterize the process. A list of the most important of these combined phenomena includes:

(a) heat generation due to plastic work,
(b) heat generation due to friction forces,
(c) accumulation of the deformation work connected with an increase of dislocation density,
(d) thermal events connected with metallurgical transformations,
(e) cooling in air or by water spray on free surfaces,
(f) cooling due to contact with the roll.

It is observed that in all of these forces, displacements and their derivatives interact with the thermally activated terms in real-time.

It is of course well known that during hot rolling the metallurgical and mechanical properties of the metal depend strongly on the temperature. Significant nonuniformity of deformation and intensive cooling by contact with the rolls lead, in most rolling processes, to sharp temperature gradients across the slab or the strip and, in consequence, to nonuniform structure and properties. Phase transformations are additional temperature dependent factors which influence the rolling process since both resistance to deformation and the grain size depend strongly on them.

The significance of the parameters and variables discussed above, their importance and their contribution to the quality of the final product have been recognized by researchers for quite some time. Monitoring the changes during the pass has been the subject of intense research, both experimental and analytical. The first and most important variable is, naturally, the temperature and the determination of its distribution in the rolled strip has been considered in numerous publications.

Most analytical models are based on the Fourier equation and it is observed that the sophistication of various solutions was stimulated by the development of computer techniques. The scope of the published work ranges from highly simplified approaches yielding approximate closed-form solutions such as the work of Lee et al. (1963), Zheleznov et al. (1968) and Seredynski (1973) to complex numerical methods giving more accurate predictions. Willmotte et al. (1973) calculated the temperature distribution in a strip using the finite difference method, assuming a uniform increase of temperature in the roll gap and employing Carslaw and Jaeger's (1959) equation to calculate cooling during contact with the work rolls. Bryant and Heselton (1982) carried out similar calculations, neglecting the effect of friction and assuming uniform heat for plastic deformation and uniform temperature across the width of the strip. Bryant and Chiu (1982a,1982b) developed models for roll temperature calculations in two-high and four-high mills. A more sophisticated solution was presented by Lahoti et al. (1978) who also applied finite difference equations of heat transfer, taking into account nonuniform heat generation in the deformation zone. They determined the strain rate field from a simplified velocity field assuming uniform strain distribution across the strip. Further, they included the temperature rise of the work roll in the computations. A similar solution with a uniform strain rate distribution across the thickness was presented by Sheppard and Wright (1980). A still more complex model, describing the thermal behaviour of the strip and the roll simultaneously was developed by Tseng (1984). He presented numerical solutions for cold and hot rolling, using as input for the heat generated in the deformation zone data obtained from direct measurements of power. Several other models of no less significance have been given by Sigalla (1957), Weber (1973), Pavlossoglou (1981a,1981b), Cerni et al. (1963) and Sellars (1985).

The development of computational techniques allowed the finite element method to simulate both thermal and mechanical events in the deformation zone. The first application of the finite element method to analyze the temperature distribution in a strip during hot rolling appears to be due to Zienkiewicz et al. (1981). Further solutions are presented by Thompson and Berman (1984), Silvonen et al. (1987), Pietrzyk and Lenard (1988) and Bertrand-Corsini et al. (1988).

Information is more sketchy as far as studies of the temperature variations during cold rolling are concerned. Roberts (1978) presented simple closed-form formulae which allow the calculation of temperature changes during cold rolling in tandem mills. Avitzur and Nowakowski (1980) applied the upper bound method and assumed adiabatic conditions to evaluate the distribution of temperatures. Their work was continued by Turczyn (1981) and Turczyn and Nowakowski (1985). Turczyn (1981) also conducted a wide experimental investigation of the temperature increase in the cold rolling process. He used the calorimetric method to measure the temperature increase due to plastic work. Poplawski and Seccombe (1980) developed a three part model of heat transfer in the cold rolling process. They used the finite difference method to determine temperature distribution in the strip and the finite element method for temperature calculations in the work roll and in the back-up roll. They verified the results of their calculations by comparison with surface temperature measurements obtained using an infrared camera. A coupled finite element simulation of cold rolling was presented by Kumar and Singhal (1987); however, their publication lacks detailed information about boundary conditions in the mechanical part of the model.

The thermal-mechanical model developed by Pietrzyk and Lenard (1988,1989a) is described in detail below. The model contains two parts. The first part determines the velocity field, the strain rate and the strain fields, using the rigid-plastic finite element approach and the second is connected with heat transfer and heat generation within the strip.

4.1 Rigid-Plastic Finite-Element Method

The elastic-plastic finite element method described in the publications of Gallagher (1975), Zienkiewicz (1977) and Owen and Hinton (1980) has been found useful for obtaining approximate solutions to problems whose exact solutions are difficult to derive. In the field of plasticity and specifically metal-forming, the method is capable of following the deformation from moment to moment, treating the material as an elastic-plastic solid. The usefulness of the method for detailed studies of the characteristics of the deformation in the rolling process was demonstrated by Thompson (1982), Liu et al. (1985,1988), Buessler and Schoenberger (1987) and by Shimazaki and Shiojima (1987). For unrestricted plastic flow, the rigid-plastic approximation is allowable locally and overall as long as the intention is to apply the solution to regions where the plastic strains are very large. Lee and Kobayashi (1973) developed a general matrix method based on one of the variational principles for rigid-plastic deformation. The method is particularly suited for problems involving large plastic deformation. The accuracy and convergence of the method were shown by Lee and Kobayashi (1973) to be excellent and it possesses all the advantageous features of the finite element technique.

The rigid-plastic finite element approach has been successfully applied to the simulation of such forming processes as upsetting (Matsumoto et al., 1977) and drawing and extrusion (Chen and Kobayashi, 1978). Due to the existence of the neutral point the solution for the rolling process was more difficult to obtain. Introduction of velocity-dependent friction stresses eased those problems considerably and several solutions incorporating two-dimensional (Li and Kobayashi, 1982; Mori et al., 1982; Pietrzyk, 1982 and Hwu and Lenard, 1988) or three-dimensional analyses (Li and Kobayashi, 1984 and Mori and Osakada, 1984) were published recently.

4.1.1 Description of the Method

The approach is based on an extremum principle which states that for a plastically deforming body of volume V under traction $\underline{F}$ prescribed on a part of the surface S and the velocity $\underline{v}$ prescribed on the remainder of the surface, the actual solution minimizes the functional

$$J = \int_V \sigma_i \dot{\epsilon}_i dV - \int_S \underline{F}\,\underline{v}\, dS \tag{4.1}$$

under the constraint $\dot{\epsilon}_V = \dot{\epsilon}_x + \dot{\epsilon}_y = 0$

where

$\dot{\epsilon}_i = \sqrt{\frac{2}{3}\dot{\underline{\epsilon}}^T \mathrm{E}\dot{\underline{\epsilon}}}$ is the effective strain rate,

$\dot{\underline{\epsilon}} = \{\dot{\epsilon}_x, \dot{\epsilon}_y, \dot{\epsilon}_{xy}\}^T$ is the vector of the strain rate components,

σ_i is the effective stress and

E is the identity matrix.

The material is assumed to be rigid-plastic and it obeys the Huber-Mises yield criterion and its associated flow rulc

$$\underline{\sigma} = \frac{2\sigma_i}{3\dot{\epsilon}_i}\mathrm{E}\dot{\underline{\epsilon}} \tag{4.2}$$

It is convenient to remove the incompressibility constraint by introducing a Lagrange multiplier λ and a stationary value problem is then obtained for the finite element analysis. The functional J then becomes

$$J = \int_V \sigma_i \dot{\epsilon}_i dV + \lambda \int_V \dot{\epsilon}_V dV - \int_S \underline{F}\,\underline{v}\, dS \qquad (4.3)$$

The discretization of the functional (4.3) follows the normal procedure in the finite element method. The basic formulae include:

- a relationship between the velocity field inside the element and the nodal velocities

$$\underline{v} = \{v_x, v_y\}^T = \mathbf{N}\,\mathbf{v} \qquad (4.4)$$

where v_x, v_y are the velocity components, $\mathbf{N}$ is the matrix of the shape functions and $\mathbf{v}$ is the vector of nodal velocities, and

- a relationship between the strain rate field inside the element and the nodal velocities

$$\underline{\dot{\epsilon}} = \{\dot{\epsilon}_x, \dot{\epsilon}_y, \dot{\epsilon}_{xy}\}^T = \mathbf{B}\,\mathbf{v} \qquad (4.5)$$

where $\dot{\epsilon}_x, \dot{\epsilon}_y, \dot{\epsilon}_{xy}$ are components of the strain rate tensor and $\mathbf{B}$ is the matrix of the derivatives of the shape functions.

Introducing equations (4.4) and (4.5) into the functional (4.3) gives

$$J = \int_V \sigma_i \sqrt{\frac{2}{3}\mathbf{v}^T \mathbf{K}\,\mathbf{v}}\, dV + \lambda \mathbf{Q}^T \mathbf{v} - \mathbf{F}^T \mathbf{v} \qquad (4.6)$$

where

$$\mathbf{K} = \mathbf{B}^T \mathbf{E} \mathbf{B}$$

$$\mathbf{Q} = \int_V \mathbf{B}^T \mathbf{C}\, dV$$

$$\mathbf{C}^T = \{1, 1, 0\}$$

$$\mathbf{F} = \int_S \mathbf{N}^T \underline{F}\, dS$$

Differentiation of equation (4.6) with respect to the nodal velocities and to the Lagrange multiplier yields

$$\frac{\partial J}{\partial \mathbf{v}} = \int_V \frac{\frac{2}{3}\sigma_i \mathbf{K}\mathbf{v}}{\sqrt{\frac{2}{3}\mathbf{v}^T\mathbf{K}\mathbf{v}}} dV + \lambda \mathbf{Q} - \mathbf{F} \tag{4.7}$$

and

$$\frac{\partial J}{\partial \lambda} = \mathbf{Q}^T \mathbf{v} \tag{4.8}$$

Relationships (4.7) and (4.8) form a set of nonlinear equations with nodal velocities $\mathbf{v}$ and Lagrange multiplier λ as the unknowns. The equations are linearized by the Newton-Raphson technique and a set of linear equations is obtained

$$\frac{\partial J}{\partial \mathbf{v}}\bigg|_{\mathbf{v}=\mathbf{v}_{i-1}} = \frac{2}{3}\int_V \frac{\sigma_i}{\dot{\epsilon}_i}\left(\mathbf{K} - \frac{2}{3}\frac{\mathbf{b}\mathbf{b}^T}{\dot{\epsilon}_i^2}\right) dV \Delta\mathbf{v} + \frac{2}{3}\int_V \frac{\sigma_i}{\dot{\epsilon}_i}\mathbf{b}\, dV + \lambda\mathbf{Q} - \mathbf{F} = 0 \tag{4.9}$$

$$\frac{\partial J}{\partial \lambda}\bigg|_{\mathbf{v}=\mathbf{v}_{i-1}} = \mathbf{Q}^T \Delta\mathbf{v} + \mathbf{Q}^T\mathbf{v} = 0 \tag{4.10}$$

where $\Delta\mathbf{v}$ is an increment of the nodal velocity and $\mathbf{b} = \mathbf{K}\mathbf{v}$.

All velocities and strain rates which appear on the right sides of equations (4.9) and (4.10) refer to the previous iteration. Iteration indices are omitted on these sides for clarity of the formulae. Left sides of equations (4.9) and (4.10) represent the derivatives of the power functional with respect to the nodal velocities and the Lagrange multiplier, calculated for the present iteration. Relationships (4.9) and (4.10) form a set of linear equations with the increments of nodal velocities $\Delta\mathbf{v}$ and the Lagrange multiplier λ as the unknowns. Expressed in matrix form, these equations are written as

$$\mathbf{P}\begin{Bmatrix}\Delta\mathbf{v}\\ \lambda\end{Bmatrix} = \mathbf{H} \tag{4.11}$$

Components of matrices H and P for two dimensional problems and for 4 node quadrilateral elements are

$$P_{kl} = \frac{2}{3}\int_V \frac{\sigma_i}{\dot{\epsilon}_i}\left(K_{kl} - \frac{2}{3}\frac{b_l b_k}{\dot{\epsilon}_i^2} \right) dV \qquad \text{for} \quad \begin{matrix} k = 1,2,\ldots,8 \\ l = 1,2,\ldots,8 \end{matrix}$$

$$P_{k9} = P_{9k} = Q_k \qquad \text{for} \quad k = 1,2,\ldots,8$$

$$P_{99} = 0$$

$$H_k = F_k - \frac{2}{3}\int_V \frac{\sigma_i}{\dot{\epsilon}_i} b_k dV \qquad \text{for} \quad k = 1,2,\ldots,8$$

$$H_9 = -\sum_{m=1}^{8} Q_m v_m$$

Matrices P and H are evaluated for each element separately and the global matrix for the whole deformation zone is assembled, as described by Gallagher (1975) and Shah and Kobayashi (1977). Subsequent steps in the finite element solution are as follows (Pietrzyk, 1983):

(a) determination of shape functions, matrices $\mathbf{B}, \mathbf{K}, \mathbf{Q}, \mathbf{P}$ and $\mathbf{H}$,
(b) inserting components of element matrices into the global matrices $\mathbf{P}$ and $\mathbf{H}$,
(c) introduction of boundary conditions which include friction stresses and tensions,
(d) elimination of equations connected with given velocities,
(e) solution of the set of linear equations by the Gauss method,
(f) introduction of the increments to the velocity field, according to the formula: $\mathbf{v}_{i+1} = \mathbf{v}_i + \xi \Delta \mathbf{v}$, where ξ is an acceleration coefficient and i represents the number of an iteration,
(g) the calculations are repeated until the solution norm

$$\rho_0 = \frac{|\Delta \mathbf{v}|}{|\mathbf{v}|} = \frac{\sqrt{\sum_{i=1}^{n} (\Delta v_{xi}^2 + \Delta v_{yi}^2)}}{\sqrt{\sum_{i=1}^{n} (v_{xi}^2 + v_{yi}^2)}} \tag{4.12}$$

is less than some assumed value. In equation (4.12) n is the number of the nodes in the finite element mesh.

Assessment of the convergence of the solution and appropriate changes of the acceleration coefficient at each step are very important in controlling the approach toward a satisfactory result. The basic criterion of convergence requires that the solution norm (4.12) decrease in each subsequent iteration. However, it has been noticed during computations that in some cases this criterion implies a divergent solution, even though it is convergent in reality. Therefore, a second criterion of convergence, measured by the error norm

$$\rho_1 = \sqrt{\sum_{i=1}^{n} \left[\left(\frac{\partial J}{\partial v_{xi}} \right)^2 + \left(\frac{\partial J}{\partial v_{yi}} \right)^2 \right] + \sum_{i=1}^{m} \left(\frac{\partial J}{\partial \lambda} \right)^2} \tag{4.13}$$

is introduced, requiring that ρ_1 decrease in subsequent iterations. This then leads to the choice of the optimum value of the acceleration coefficient in each step toward the solution. In equation (4.13) n is the number of nodes and m is the number of elements.

4.1.2 Boundary Conditions

Boundary conditions in the rolling process include friction stresses on the contact surface, the shape of that surface and tensions at the entry and exit regions. Usually, in other stationary processes such as drawing or extrusion the entry and exit velocities are also introduced as the boundary conditions. During rolling the position of the neutral point is not known *a priori* and, in consequence, the entry and exit velocities cannot be imposed as boundary conditions. That fact delayed the application of the rigid-plastic finite element technique to the simulation of rolling in comparison with other forming processes such as drawing, upsetting and extrusion. The difficulty connected with the evaluation of the location of the neutral point was overcome by the introduction of velocity dependent friction stresses. The function suggested by Chen and Kobayashi (1978b) for the ring upsetting process

$$\frac{\mu}{\mu_0} = \frac{2}{\pi}\tan^{-1}\frac{v_s}{a} \tag{4.14}$$

was first used by Li and Kobayashi (1982) to simulate friction phenomena in rolling. In equation (4.14) μ is the friction coefficient, μ_0 is a constant value of the friction coefficient, v_s is the slip velocity and a is a constant, assumed to be 10^{-3}.

The choice of the constant a is important since the magnitude of the frictional forces and the location of the neutral point are both affected by it. Its effect on the distribution of the friction coefficient is shown in Fig.4.1, where the variations of the ratio of μ/μ_0 along the arc of contact are plotted for $a = 0.001$, 0.01 and 0.1. It is observed that increasing the value of a moves the neutral point toward the exit. A limitation of the frictional model of equation (4.14) is noted when a is too large. For $a = 0.1$ the unrealistic prediction of having no forward slip zone in the roll gap is obtained. The general conclusion, that the value of a should be approximately 3 orders of magnitude smaller than the average slip velocity is the most appropriate, may therefore be arrived at. Increasing a leads to the prediction of unrealistic frictional conditions and decreasing a slows the convergence of the iterative procedure.

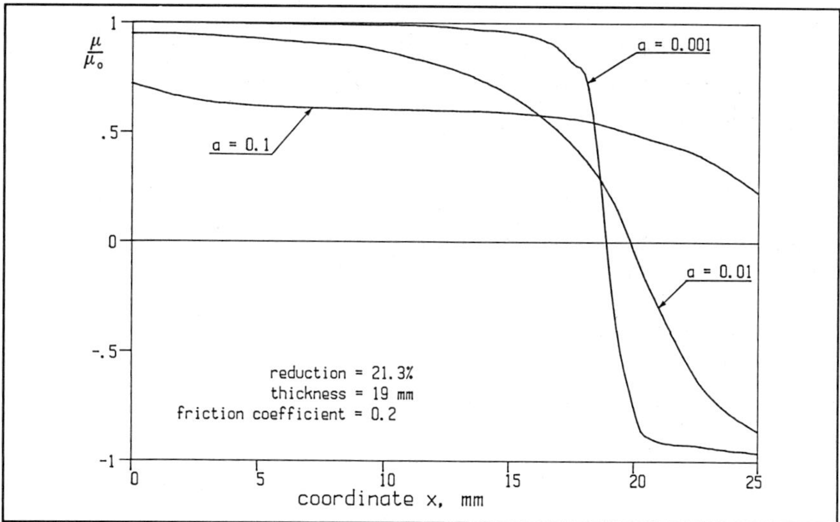

Fig.4.1 Variation of friction coefficient for various values of the parameter a.

In order to introduce the friction stresses into the matrices **P** and **H** in equation (4.11) the power of friction losses needs to be calculated. These are given by

$$J_f = \int_S \int_0^{v_s} \frac{2}{\pi} \mu_0 \sigma_p \tan^{-1}\frac{v_s}{a} dv_s dS \tag{4.15}$$

The method of the evaluation of the terms of matrices **P** and **H**, connected with friction stresses, will be presented for two adjacent nodes on the contact surface (Fig.4.2). Generally, in the boundary conditions for the rolling process stresses normal to the surface p are prescribed to be zero and stresses tangent to the surface τ are given as the friction stresses. The slip velocity between the nodes is given as a function of its components

$$v_s = \hat{\mathbf{N}} \mathbf{v}_s \tag{4.16}$$

where

$$\hat{\mathbf{N}}^T = \{\hat{N}_1, \hat{N}_1, \hat{N}_2, \hat{N}_2\}$$

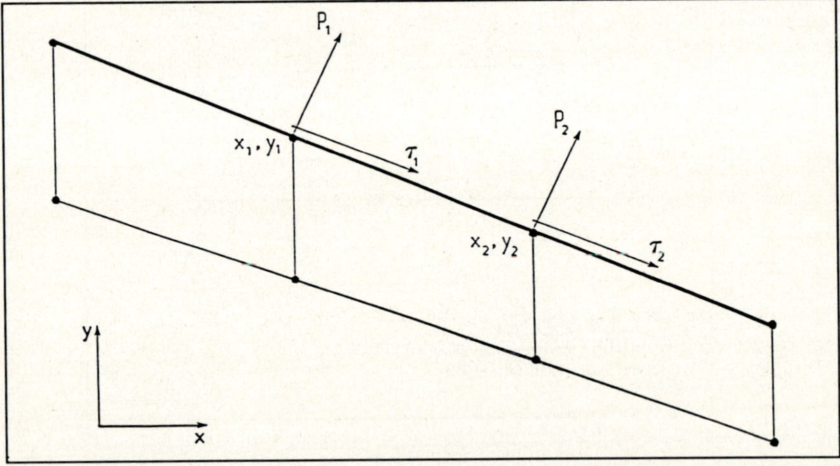

Fig.4.2 Tractions in two adjacent nodes on the contact surface.

$$\hat{N}_1 = \frac{x - x_2}{x_1 - x_2} \qquad \hat{N}_2 = \frac{x - x_1}{x_2 - x_1}$$

are the surface shape functions, and

$$\mathbf{v}_s = \begin{Bmatrix} (v_r \cos\gamma - v_{x1})\cos\gamma \\ (v_r \sin\gamma + v_{y1})\sin\gamma \\ (v_r \cos\gamma - v_{x2})\cos\gamma \\ (v_r \sin\gamma + v_{y2})\sin\gamma \end{Bmatrix}$$

In the above equation v_r represents the surface velocity of the roll. Differentiation of equation (4.15) with respect to nodal velocities $\mathbf{v}^T = \{v_{x1}, v_{y1}, v_{x2}, v_{y2}\}$ yields

$$\frac{\partial J_f}{\partial \mathbf{v}} = \frac{2}{\pi}\mu_0 \int_S \tan^{-1}\frac{v_s}{a}\mathbf{G}\hat{\mathbf{N}}\,dS \tag{4.17}$$

$$\frac{\partial^2 J_f}{\partial \mathbf{v}\partial \mathbf{v}^T} = \frac{2}{\pi a}\mu_0 \int_S \frac{\mathbf{G}\hat{\mathbf{N}}\hat{\mathbf{N}}^T\mathbf{G}^T}{1 + \left(\frac{v_s}{a}\right)^2}dS \tag{4.18}$$

where

$$\mathbf{G} = \frac{\partial \mathbf{v}_s}{\partial \mathbf{v}} = \begin{bmatrix} -\cos\gamma & 0 & 0 & 0 \\ 0 & \sin\gamma & 0 & 0 \\ 0 & 0 & -\cos\gamma & 0 \\ 0 & 0 & 0 & \sin\gamma \end{bmatrix}$$

and γ is an angle between the tangent to the roll surface and the horizontal axis. This angle varies along the arc of contact, decreasing from its initial value which is equal to the angle of bite at the entry to zero at the exit.

Introducing equations (4.17) and (4.18) into (4.11), the final formulae for matrices $\mathbf{P}$ and $\mathbf{H}$ are obtained

$$P_{kl} = \frac{2}{3}\int_V \frac{\sigma_i}{\dot{\epsilon}_i}\left(K_{kl} - \frac{2b_l b_k}{3\,\dot{\epsilon}_i^2}\right)dV + \frac{2}{\pi a}\mu_0 \int_S \frac{\sum_{i=1}^{8} G_{li}\hat{N}_i \sum_{i=1}^{8} G_{ki}\hat{N}_i}{1+\left(\frac{v_s}{a}\right)^2}dS \qquad \begin{array}{l} k=1,2,\ldots,8 \\ l=1,2,\ldots,8 \end{array}$$

$$P_{k9} = P_{9k} = Q_k \qquad \text{for} \qquad k=1,2,\ldots,8$$

$$P_{99} = 0$$

$$H_k = F_k - \frac{2}{3}\int_V \frac{\sigma_i}{\dot{\epsilon}_i} b_k dV + \frac{2}{\pi}\mu_0 \int_S \tan^{-1}\frac{v_s}{a}\sum_{i=1}^{8} G_{ki}\hat{N}_i dS \qquad \text{for} \qquad k=1,2,\ldots,8$$

$$H_9 = -\sum_{m=1}^{8} Q_m v_m$$

Vector F in equations (4.11) now includes all external forces except the friction forces. In single stand roughing mills F = 0, while in strip rolling the components of F are the interstand tension stresses.

The finite element mesh and the externally prescribed conditions used in the calculations are presented in Fig.4.3. The constraints for velocities include the conditions that $v_y = 0$ along the axis of symmetry and that $v_y = -v_x \tan\gamma$ on the contact surface. Tension tractions $\sigma_x = 0$ for rolling without tensions or $\sigma_x = \sigma_1$ and $\sigma_x = \sigma_2$ for rolling with tensions are prescribed on the entry and exit surfaces. Friction forces $\tau_x = \mu\sigma_p\cos\gamma$ and $\tau_y = \mu\sigma_p\sin\gamma$ are introduced at all nodes on the contact surface by equations (4.17) and (4.18).

Special attention is paid to the modelling of the shape of the free surfaces on both sides of the sample in front of the roll gap. Usually, conditions of zero traction (F = 0) and zero mass flow ($\mathbf{v} \cdot \mathbf{n} = 0$) have to be satisfied on the free surfaces. However, in modelling of the steady state processes only the first condition is imposed by the boundary conditions. As far as the condition of zero mass flow is considered, it is usually assumed to be satisfied automatically for the mesh being used. It is observed during calculations however, that zero normal velocity is not always obtained on the top and bottom free surfaces, in particular when modelling rolling processes with large shape coefficients. This creates some problems with the convergence of the solution and with the accuracy of the results. In order to overcome this difficulty the shape of the free surface of the sample is corrected during calculations in the present model. Corrections are introduced according to the suggestion made by Dawson and Thompson (1978), who assumed that when a segment of a boundary is a free surface, the control volume must be adjusted to make a boundary a stream line. This is accomplished by using numerical integration and equating the slope of the free surface to the ratio of the

becomes a two point boundary value problem. In general, however, the iterative procedure does not converge to an assumed initial point of contact between the roll and the sample and the position of this point has to be corrected during calculations.

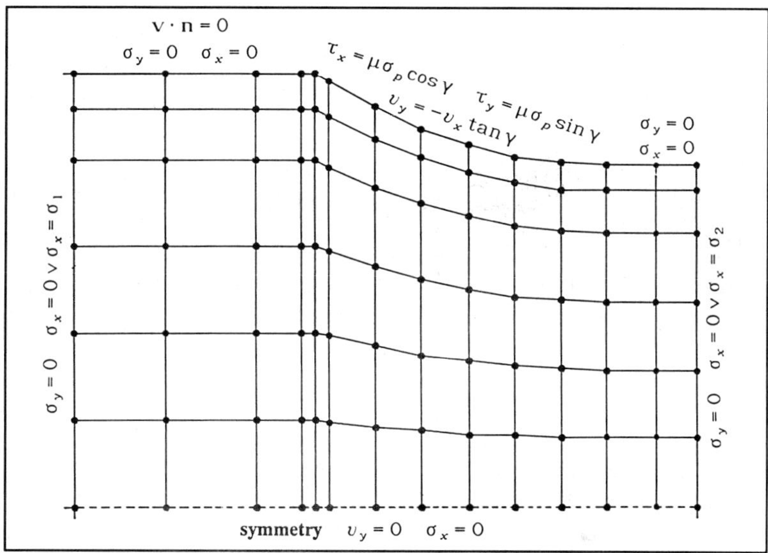

Fig.4.3 Finite element mesh and externally prescribed conditions.

Introduction of the correction of the shape of the free top surface improves the convergence of the program significantly. Moreover, it avoids the unrealistic increase of the predicted shear strain rates in the element close to the initial point of contact, which is particularly significant for large shape coefficients, thus, for large angles of bite. Correction of the free surface results in a smooth shape of the boundary close to the initial point of contact.

4.1.3 Structure of the Computer Program

The structure of the computer program for the rigid-plastic finite-element model is identical to that used earlier by Chen and Kobayashi (1978a) for axisymmetrical drawing or extrusion and by Matsumoto and Kobayashi (1977) for compression and is described by Pietrzyk (1983a) and Pietrzyk and Glowacki (1985). The main program is used to determine the required dimensions of arrays and it prevents excessive use of computer memory. It also calls the subroutines READ and PLAST. All data involving process parameters, material properties and mesh dimensions are introduced in READ and all finite element calculations are performed in PLAST whose flow chart is presented in Fig.4.4. The following subroutines are used during calculations:

- MSTIFF - calculates the stiffness matrix and the strain rate field,
- ELGAUSS - solves a set of linear equations $AX+B=0$ for band matrix A,
- STRAIN - calculates the flow lines and strains by integration of the strain rates along the flow lines,
- HARD - calculates the yield stress of the material,
- MODIFY - constrains the stiffness matrix by elimination of equations for which the velocities are known,
- FORCE - calculates the forces at the nodes on the contact surface.

This structure of the program guarantees low memory requirements and computational time.

4.2 Heat Transfer

4.2.1 Numerical Solution of the Fourier Equation

The general diffusion equation for a two-dimensional problem is

$$\nabla^T (k \nabla T) + Q = \rho c_p \frac{\partial T}{\partial t} \tag{4.19}$$

where T is the temperature, Q is the rate of heat generation due to plastic deformation, k is the heat conduction coefficient, c_p designates the specific heat and ρ is the density.

Solution of equation (4.19) has to satisfy the specified boundary conditions. There are three commonly used types of boundary conditions which are applicable to the simulation of forming processes, see for example Ozisik (1985):

(a) the temperature is prescribed along the boundary surface and, in a general case, it is a function of both time and position,

(b) the normal derivative of the temperature is prescribed at the boundary surface, and it may be a function of both time and position,

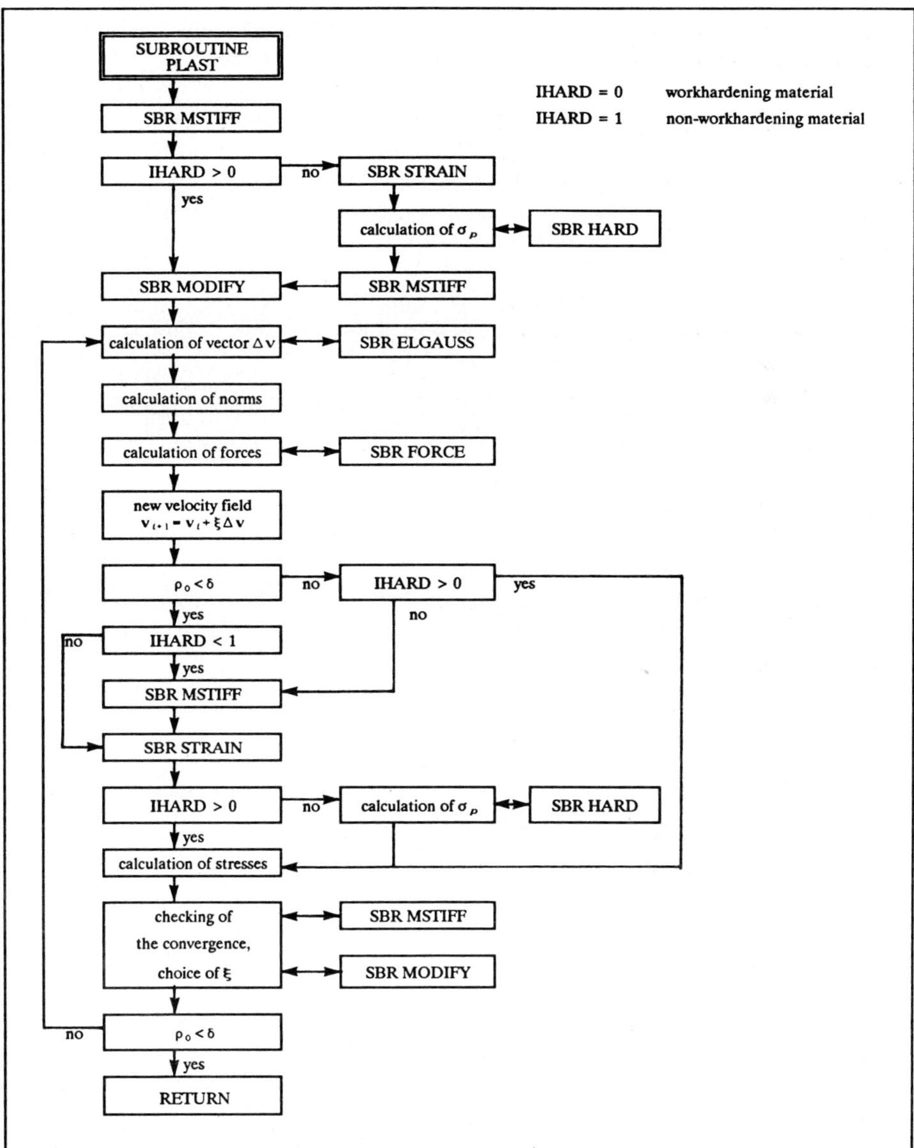

Fig.4.4 Flow chart of the subroutine PLAST.

(c) a linear combination of the temperature and its normal derivative is prescribed at the boundary surface.

Generally, the last two kinds of boundary conditions are useful in modelling the rolling processes. The physical meaning of these conditions is that the heat flux through the contact surface is represented by heat generated due to the friction losses and by heat conducted to the roll or by heat dissipated by convection and radiation. To demonstrate this the energy balance for the boundary surface is written

$$k\frac{\partial T}{\partial n} = \alpha(T_0 - T) + q \tag{4.20}$$

where n is a coordinate normal to the surface, α is the heat transfer coefficient, T_0 is the roll temperature or ambient temperature and q is the rate of heat generation due to the friction forces.

Two distinct procedures are available for obtaining an approximate solution of equation (4.19). The first is the method of weighted residuals while the second uses variational principles. Both of these are described in detail by Zienkiewicz (1977). The latter method states that a minimum of the general functional

$$J = \int_V F\left(x, y, \frac{\partial T}{\partial x}, \frac{\partial T}{\partial y}, T\right) dV + \int_S E\left(x, y, \frac{\partial T}{\partial x}, \frac{\partial T}{\partial y}, T\right) dS \tag{4.21}$$

is obtained when the temperature field $T(x, y)$ satisfies the Euler's equations

$$\begin{aligned} &\frac{\partial F}{\partial T} - \nabla\left[\frac{\partial F}{\partial(\nabla T)}\right] = 0 \qquad \text{in} \quad V \\ &\frac{\partial E}{\partial T} - \nabla\left[\frac{\partial E}{\partial(\nabla T)}\right] = 0 \qquad \text{on} \quad S \end{aligned} \tag{4.22}$$

The solution of equation (4.19) then reduces to searching for a functional for which the minimization yields Euler's equations. It is quite straightforward to show that the functional

$$J = \int_V \left\{\frac{k}{2}\left[\left(\frac{\partial T}{\partial x}\right)^2 + \left(\frac{\partial T}{\partial y}\right)^2\right] - \bar{Q}T\right\} dV - \int_S \left[\alpha\left(T_0 - \frac{T}{2}\right)T + qT\right] dS \tag{4.23}$$

gives on minimization the satisfaction of the original problem set in equation (4.19) as well as the boundary conditions given by (4.20). The algebraic manipulations verifying the above are presented by Zienkiewicz (1977). In equation (4.23) $\overline{Q}$ is given by

$$\overline{Q} = Q - c_p \rho \frac{\partial T}{\partial t}$$

Discretization is performed in the usual finite element manner. The temperature inside an element is presented as a function of nodal values according to the following formula

$$T = \Sigma N_i T_i = \mathbf{N}\mathbf{T} \tag{4.24}$$

where $\mathbf{T}$ is a vector of nodal temperatures, T_i represents components of the vector of nodal temperatures, $\mathbf{N}$ is a vector of shape functions and N_i represents the shape functions.

Substitution of the relationship (4.24) into equation (4.23) yields

$$J = \int_V \left\{ \frac{k}{2} \left[\left(\Sigma \frac{\partial N_i}{\partial x} T_i \right)^2 + \left(\Sigma \frac{\partial N_i}{\partial y} T_i \right)^2 \right] - \overline{Q} \Sigma N_i T_i \right\} dV$$

$$- \int_S \left[\alpha \left(T_0 - \frac{1}{2} \Sigma N_i T_i \right) \Sigma N_i T_i + q \Sigma N_i T_i \right] dS \tag{4.25}$$

Minimization of the functional (4.25) requires calculation of the partial derivatives with respect to nodal temperatures and it results in the following set of linear equations

$$\frac{\partial J}{\partial T_j} = \int_V k \left[\Sigma \left(\frac{\partial N_i}{\partial x} T_i \right) \frac{\partial N_j}{\partial x} + \Sigma \left(\frac{\partial N_i}{\partial y} T_i \right) \frac{\partial N_j}{\partial y} \right] dV - \int_V \overline{Q} N_j dV$$

$$- \int_S (\alpha T_0 + q) N_j dS + \int_S \alpha \Sigma (N_i T_i) N_j dS \tag{4.26}$$

Equations (4.26) written in a matrix form are

$$\mathbf{H}\mathbf{T} + \mathbf{P} = 0 \tag{4.27}$$

where

$$H_{ij} = \int_V (\nabla N_i)^T k (\nabla N_j) dV + \int_S \alpha N_i N_j dS$$

and

$$P_i = -\int_S (\alpha T_0 + q) N_i dS - \int_V \bar{Q} N_i dV$$

Since the rate of heat generation due to the plastic work Q and the heat generated due to the friction losses q are generally given in the form of nodal values, it is often more convenient to write the vector **P** as

$$P_i = \int_S [\alpha T_0 N_i + \Sigma(N_j q_j) N_i] dS - \int_V \Sigma(N_j \bar{Q}_j) N_i dV \tag{4.28}$$

Equation (4.27) consists of a set of linear equations, solution of which gives the values of the nodal temperatures T_i during a stationary thermal state when $\partial T / \partial t = 0$ and $\bar{Q} = Q$. However, nonstationary heat transfer problems are sometimes involved in the modelling of hot and cold rolling. During finite but reasonably short time intervals the partial derivatives of the temperature can be considered as functions of the x and y coordinates only. Then the solution of equation (4.19) is obtained as described, from

$$\mathbf{HT} + \mathbf{C}^T \frac{\partial}{\partial t} \mathbf{T} + \mathbf{P} = 0 \tag{4.29}$$

where

$$C_{ij} = \int_V N_i c_p \rho N_j dV \tag{4.30}$$

The assumption that the nodal temperatures are linear with respect to time leads to

$$\mathbf{T} = \{\mathbf{N}_0, \mathbf{N}_1\} \begin{Bmatrix} \mathbf{T}_i \\ \mathbf{T}_{i+1} \end{Bmatrix} \tag{4.31}$$

where

$$N_0 = \frac{\Delta t - t}{\Delta t} \quad ; \quad N_1 = \frac{t}{\Delta t}$$

and Δt designates a time interval, t is the time and T_i and T_{i+1} are the nodal values of temperature for $t = 0$ and $t = \Delta t$, respectively. Using (4.31), the derivative of the temperature with respect to time can be expressed as

$$\frac{\partial T}{\partial t} = \left\{ \frac{\partial N_0}{\partial t}, \frac{\partial N_1}{\partial t} \right\} \left\{ \begin{matrix} T_i \\ T_{i+1} \end{matrix} \right\} = \frac{1}{\Delta t} \{-1, 1\} \left\{ \begin{matrix} T_i \\ T_{i+1} \end{matrix} \right\} \tag{4.32}$$

and since the vector of nodal temperatures T_i is known, only one weighted residual is required to allow integration of equation (4.30) with respect to time. The integral to be evaluated then becomes

$$\int_0^{\Delta t} \frac{t}{\Delta t} \left[H\{N_0, N_1\} \left\{ \begin{matrix} T_i \\ T_{i+1} \end{matrix} \right\} + C \left\{ \frac{\partial N_0}{\partial t}, \frac{\partial N_1}{\partial t} \right\} \left\{ \begin{matrix} T_i \\ T_{i+1} \end{matrix} \right\} + P \right] dt = 0 \tag{4.33}$$

Introduction of the shape functions, given by equation (4.31), yields

$$\int_0^{\Delta t} \frac{t}{\Delta t} \left[\left(\frac{\Delta t - t}{\Delta t} T_i + \frac{t}{\Delta t} T_{i+1} \right) H + (-T_i + T_{i+1}) \frac{C}{\Delta t} + P \right] dt = 0 \tag{4.34}$$

Rearrangement of (4.34) gives the final equation which is linear with respect to T_{i+1}, as follows

$$\left(2H + \frac{3}{\Delta t} C \right) T_{i+1} = \left(-H + \frac{3}{\Delta t} C \right) T_i - 3P \tag{4.35}$$

Equation (4.35) allows calculation of the vector of nodal temperatures T_{i+1} after a time interval Δt, provided that the initial nodal temperatures at $t = 0$ are known. The non-steady state model should be applied to simulate the temperature distribution during reverse rolling of ingots or billets, when the effect of cooling of the ends is significant. An example of an application of the non-steady state model to the simulation of three-dimensional temperature fields during rolling of rectangular slabs is presented by Pietrzyk and Lenard (1988).

4.2.2 Steady-State Model with Convection

In the majority of the rolling processes, including all continuous processes, steady-state conditions can be assumed to exist but the convective term then must be introduced in equation (4.19) to account for the mass flow. A typical convective-diffusion equation for a steady-state problem is of the form

$$\nabla^T(k\nabla T) + Q - \rho c_p \mathbf{v}^T \nabla T = 0 \tag{4.36}$$

Serious difficulties can be encountered when solving the equation (4.36). First, the convective acceleration term introduces a non-symmetric matrix into the final equation. Further problems arise when the standard processes of discretization are applied. According to Zienkiewicz (1977), if the Galerkin-Bubnov weighted residual method is used, an oscillatory solution will be found with the local Peclet number

$$\mathrm{Pe} = \frac{\max(|v_x|, |v_y|)\rho c_p d}{k} \tag{4.37}$$

exceeding a value of 2. In the above the mesh size is indicated by d and v_x and v_y are the velocity components. Application of the weighted residual method to equation (4.36) yields

$$\int_V \nabla^T w_i k \nabla T dV - \int_V \rho c_p w_i \left(v_x \frac{\partial T}{\partial x} + v_y \frac{\partial T}{\partial y} \right) dV$$
$$- \int_V w_i Q dV - \int_S w_i(\alpha T_0 + q) dS + \int_S \alpha w_i T dS = 0 \tag{4.38}$$

where w is a weighting function.

The subsequent substitution of (4.24) and selection of a suitable shape function produces the classical system of linear discretized equations

$$\mathbf{HT} + \mathbf{P} = 0 \tag{4.39}$$

where

$$H_{ij} = \int_V (\nabla^T w_i k \nabla N_j + \rho c_p w_i \mathbf{v}^T \nabla N_j) dV + \int_S \alpha w_i N_j dS$$

$$P_i = -\int_V w_i Q dV - \int_S w_i (\alpha T_0 + q) dS$$

Now the formulations and solutions can be obtained using any desired element form or weighting function. As mentioned above however, application of the standard Galerkin procedure, which assumes that $w_i = N_i$, leads to meaningless oscillations in all cases when the convective terms are significant. This problem is usually encountered in the simulation of continuous rolling processes, where rolling velocities often exceed 10 m/s and the local Peclet number is much greater than 2. The difficulty just mentioned had been pointed out in several publications on finite element simulation of convective transport problems, e.g. Zienkiewicz (1977), Heinrich and Zienkiewicz (1977) and Heinrich et al. (1977) where it is shown that if a non-symmetric weighting function is used in place of the standard shape function then it is possible not only to avoid oscillations but also to improve the accuracy of the solution. The main features of such upwinded functions can be easily described for a one-dimensional problem with basic shape functions of the linear kind. The upwinded weighting function for this case is

$$w_i = N_i - \beta f(x) \tag{4.40}$$

where $f(x)$ is some positive function, such that $f(x)$ is zero at the nodes and

$$\int_0^d f(x) dx = \frac{1}{2} d$$

for each element. In equation (4.40) β is a constant which is positive if the velocity is towards the node and negative if the velocity is away from the node. Details of such upwinded weighting functions for two-dimensional and three-dimensional elements are fully described by Heinrich et al. (1977), Heinrich and Zienkiewicz (1977) and Zienkiewicz (1977). Some general observations, indicating the best choice of the coefficient β in equation (4.40), are given by Zienkiewicz (1977). Introduction of upwinded weighting functions in equation (4.39) facilitates steady-state solutions for the rolling process even for the high rolling velocities, usually encountered in continuous hot strip mills.

The steady-state model can also be applied to simulate the temperature rise in the roll during rolling. Derivation of equations in that case follows the same procedure as for the strip temperatures, only the deformation heating Q needs to be eliminated from the heat flux formula. Finally, the same set of linear equations (4.39) is obtained for the roll temperature with the vector **P** given as

$$P_i = -\int_S w_i(\alpha T_0 + q)\,dS \tag{4.41}$$

In equation (4.41) T_0 represents the strip temperature or the ambient temperature.

The solution of both equation (4.35) and (4.39) requires additional data connected with the rates of heat generation. These are supplied by the second component of the model which is concerned with the determination of the velocity, strain and strain rate distributions. The method suggested by Lee and Kobayashi (1973) is used here in formulating the finite element solution for the rigid-plastic material. Detailed description of this method is presented in Section 4.1. The technique allows the determination of the plastic work per unit volume and time as a product of the effective stress σ_i and the effective strain rate $\dot{\epsilon}_i$.

The rate of heat generation Q should take into account both positive and negative effects of dislocation density as described by Rebelo and Kobayashi (1980) and is given by equation (1.11). This equation should be introduced in all models which deal with cold and warm rolling processes. In the hot rolling process, however, the contribution of these extra terms to heat generation is small when compared to other effects and the material constants which appear in the equation (1.11) are difficult to determine. Therefore, in the present model all of the work of plastic deformation in hot rolling is assumed to be converted into heat. Of course, the strain hardening and softening phenomena are to be taken into consideration in the flow strength calculations.

Both functionals for the heat transfer and for the velocity field should be solved simultaneously. In the present model the temperature field is calculated for each iteration during the optimization of functional (4.3) in all hot rolling calculations, where the mechanical properties are temperature sensitive. Current velocity, strain, strain rate and stress fields are used in the temperature calculations. The temperature field is used to determine the material's resistance to deformation in each element. In all cold rolling processes, where the temperature dependence of mechanical properties is weak or negligible, the temperature calculations are performed for every few iterations during optimization of the power functional.

4.3 Results of Computations

The model described in the previous sections is used to calculate the temperature distribution as well as the velocity, strain rate, strain and stress fields in the deformation zone during hot and cold rolling of various materials. Plane strain state is assumed in the velocity field calculations and four node quadrilateral elements are used. The finite element mesh and boundary conditions for the mechanical part of the model are presented in Fig.4.3. The formulation for the thermal component differs for the rolling of square sections and for the flat rolling processes. During hot rolling of slabs with the cross-section close to square the heat transfer in the transverse direction is of significant importance and it cannot be neglected. Therefore, calculations are performed on a mesh consisting of 12-node quadrilateral Serendipity family elements which is located in a plane perpendicular to the slab axis and is allowed to move with the actual slab velocity. The non-steady temperature state is obtained in this plane using equation (4.35). The initial shape of the finite element mesh and thermal boundary conditions are shown in Fig.4.5. In continuous hot and cold rolling processes conduction in the transverse direction can be neglected and the finite element mesh of 12-node quadrilateral Serendipity family elements is then located in the plane parallel to the strip axis. In the corners

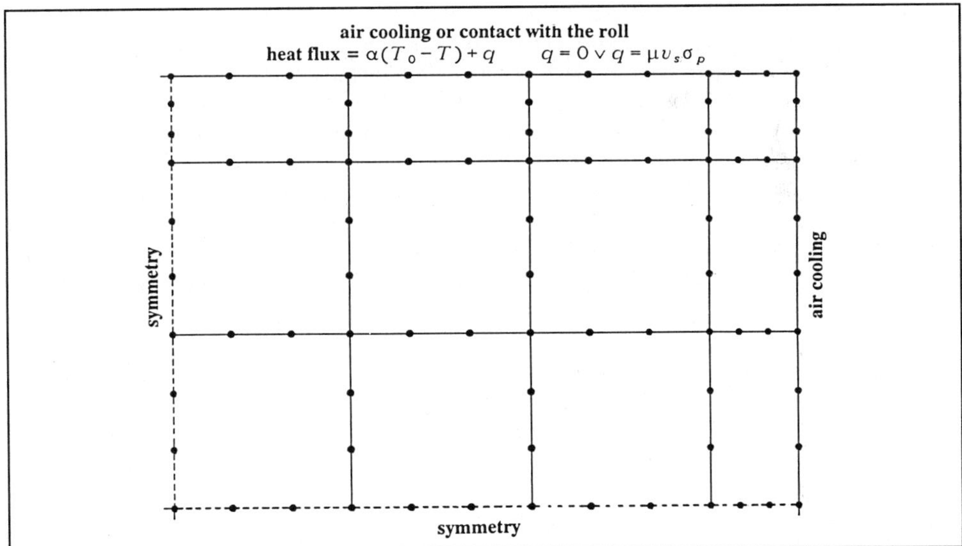

Fig.4.5 Finite element mesh and boundary conditions; nonsteady-state temperature model.

it overlaps with the mesh of the mechanical component of the model, which is presented in Fig.4.3. Steady-state solution, described by equation (4.39), is applied in these cases of rolling. The temperature is specified at the entrance to the control volume, while the plane of symmetry and the exit plane are treated as insulated surfaces. The temperature across the roll-workpiece interface is allowed to be discontinuous to account for the discontinuity created by the relative velocity between the roll and the strip and by the resistance to heat transfer across the interface. This latter can be specified as a function of the normal pressure, although for the purposes of the present model it is assumed to be constant and the heat flux through the interface is taken to be proportional to the heat transfer coefficient and to the temperature difference between the surfaces in contact. The value of the heat transfer coefficient is carefully chosen, depending on the rolling conditions, on the basis of experimental data. Detailed investigation of the role of the heat transfer coefficient in the modelling of thermal events during rolling is presented in Section 4.4. The roll temperature may be assumed to remain constant or it can be calculated using the steady-state finite-element model represented by equation (4.39) with the matrix P given by equation (4.41). The boundary conditions for the roll are similar to those for the strip. The temperature is prescribed on the entry surface while the exit plane from the control volume and the boundary inside the roll are treated as insulated surfaces. Heat transfer through the contact surface is also modelled by means of the heat transfer coefficient and an iterative procedure is used to solve the problem for both the strip and the roll. The free surfaces on the strip and on the roll are subject to air cooling and the coefficients of convective and radiative heat transfer are specified for those surfaces.

The model enables the calculation of the distributions of velocities, strain rates, strains, stresses and temperatures in the deformation zone. All material parameters, necessary for heat transfer calculations are introduced as functions of the temperature on the basis of experimental data presented by Touloukian et al. (1970a, 1970b) and Touloukian and Buyco (1970). Flow strength in hot forming processes is introduced as a function of temperature, strain, strain rate and carbon content using Shida's (1974) equations (see also Section 2.1.1). In the analysis of the warm and cold rolling processes experimentally developed strain hardening curves are used. A compilation of the temperature dependent mechanical and thermal properties for various materials is given in Table 4.1 on page 135.

Frictional coefficients used in all calculations in this section are 0.2 for hot rolling and 0.06 for cold rolling. In order to avoid discrepancies connected with evaluation of the real time of rolling, the rolls' elastic flattening is taken into account. The length of the contact is calculated by integration of the biharmonic equation (Roychoudhuri and Lenard, 1984) - see also Section 3.2.2.

4.3.1 Hot Rolling

Typical results for hot rolling of a 10 mm thick steel sample are presented in Figs.4.6 - 4.10. The sample is assumed to be heated to a uniform temperature of 1000°C and cooled in air for 10 s. Reduction is 30% and the roll radius is taken to be 200 mm. The velocity field in the deformation zone is shown in Fig.4.6. Each line segment in this figure represents the difference between the vectors of the roll velocity and that of the plastically deforming metal. Both the

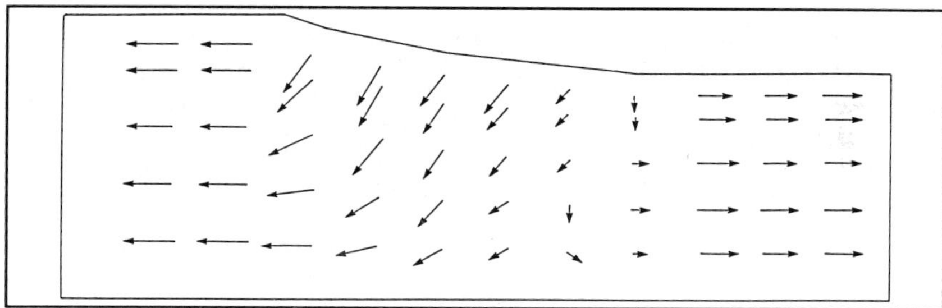

Fig.4.6 Velocity field in the roll gap.

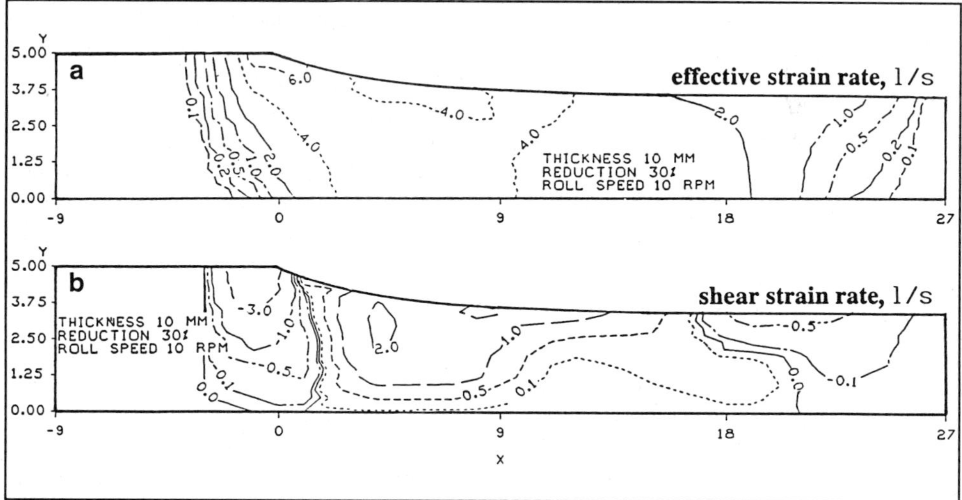

Fig.4.7 Effective strain rate (a) and shear strain rate (b) distributions during hot rolling.

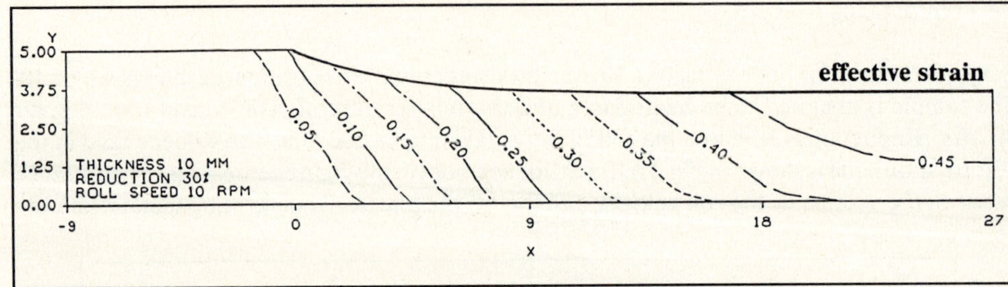

Fig.4.8 Effective strain distribution during hot rolling.

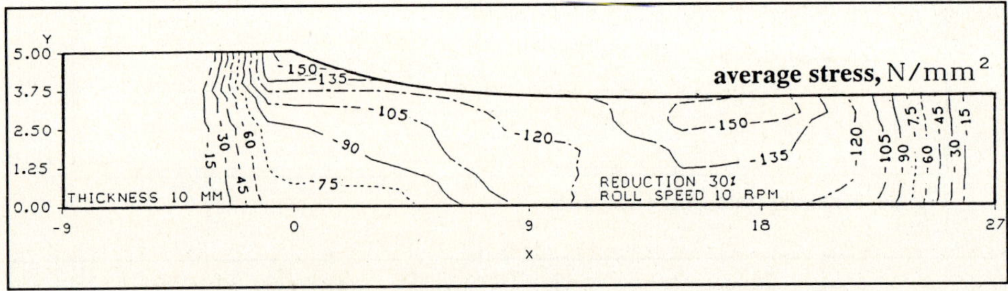

Fig.4.9 Average stress distribution during hot rolling.

direction and the value of the relative velocity can be determined from the figure, the former being represented by the direction of the line segment and the latter by its length. Effective strain rate and shear strain rate distributions are shown in Fig.4.7. Strain rate concentrations, characteristic of the rolling process, are observed near the initial point of contact between the roll and the workpiece. The distribution of the effective strain is presented in Fig.4.8. Due to the shear strain rate concentrations, the strains are distributed nonuniformly after rolling. The largest values appear close to the contact zone. The field of the average stress, defined as $\sigma_m = (\sigma_x + \sigma_y)/2$, is shown in Fig.4.9. Since rolling without tensions is considered, a compressive state of stress appears everywhere in the deformation zone. Finally, results of calculations of the temperature distributions in the strip centre and in the exit plane are shown in Fig.4.10. The heat transfer coefficient $\alpha = 4800$ W/m^{2}K, determined experimentally by Karagiozis and Lenard (1988) and Pietrzyk and Lenard (1989b) for hot rolling in laboratory conditions and with an assumed constant roll temperature of 40^{0}C, is used in the temperature calculations.

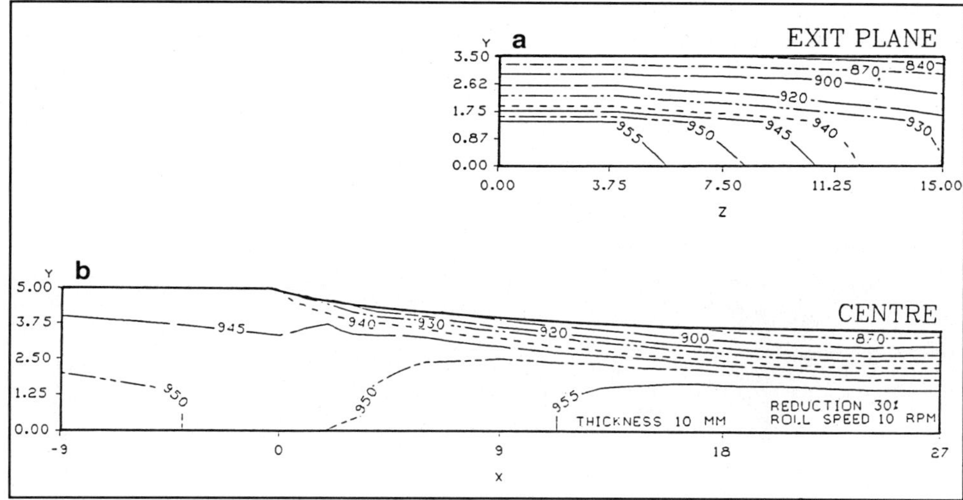

Fig.4.10 Temperature distribution during hot rolling.

The non-steady state solution for the temperature field in a plane perpendicular to the strip axis and moving with the actual strip velocity is applied here. This method of solution enables the determination of the temperature distribution in the whole deformation zone using a two-dimensional formulation, as shown by Pietrzyk and Lenard (1988). Fig.4.10b shows the lines of constant temperature on a plane, parallel to the direction of rolling. As expected, the coolest part of the strip is the one near the contact zone and sharp temperature gradients toward the strip's centre can be observed. The slow rate of heat loss of the central portion before entering the deformation zone is also noticed; recall, that the strip was assumed to be heated to a uniform temperature of 1000°C in the furnace. The same part of the strip is subjected to some heating in the roll gap, the result of work done on the plastically deforming metal. The temperature distribution in a plane perpendicular to the direction of rolling and located at exit is shown in Fig.4.10a. The hot centre and cool surface are again observed. As well, the different rates of cooling across and through the strip are also indicated.

Results of simultaneous calculations of temperature distributions in both the strip and the roll are presented in Fig.4.11. The strip thickness at entry was 19 mm, the reduction in the pass was 20% and the roll speed was 4 rpm. Intensive heating of the roll surface is observed with the maximum temperature reaching 280°C. This increase influences the boundary conditions, in particular, the value of the heat transfer coefficient, to be discussed in detail in Section 4.5. The value of $\alpha = 13000 W/m^2K$, appropriate for hot rolling of steel in laboratory conditions, is used in Fig.4.11.

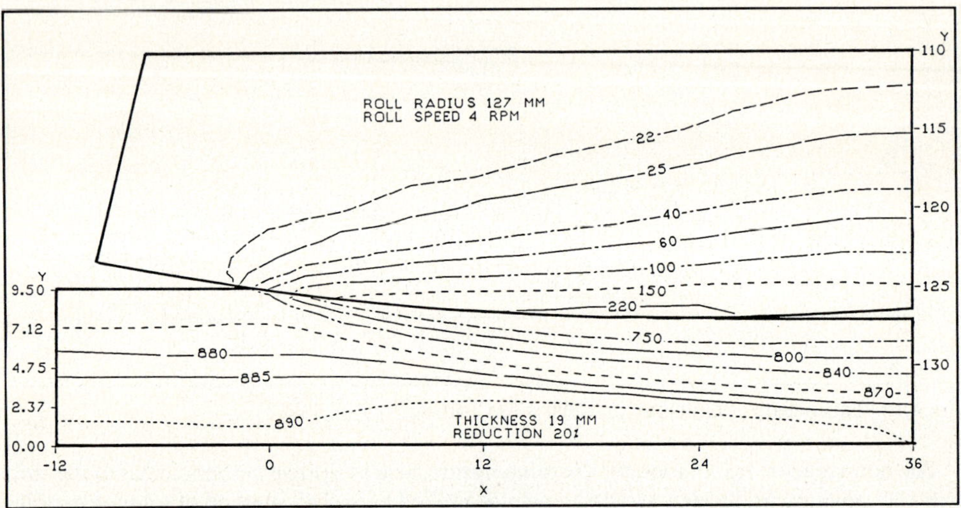

Fig.4.11 Temperature distribution in the strip and in the roll during hot rolling.

4.3.2 Cold Rolling

A similar set of results for cold rolling of a 2 mm thick low carbon steel strip is presented in Figs.4.12 - 4.15. Reduction in the pass is 20% and roll radius is taken to be 160 mm. The flow curve for the steel and all material parameters, necessary for heat transfer calculations are given in Table 4.1.

The value of the heat transfer coefficient is not expected to affect the predictions of the model for cold rolling in a significant way. Experimentally evaluated values of this coefficient for cold forming processes have not been widely reported in the literature. The value 7500 W/m^2K measured by Semiatin et al. (1987) in a two die experiment is used in the present calculations. The numerical values of the parameters in equation (1.11) are taken from Rebelo and Kobayashi (1980) as

$$a = 125 \times 10^{8} \mathrm{mm}^{-2}$$
$$b = 155 \times 10^{3} \mathrm{s}^{-1}$$
$$D/k = 17000\ \mathrm{K}$$
$$\nu = 317 \times 10^{-11} \mathrm{Nmm/mm}\ .$$

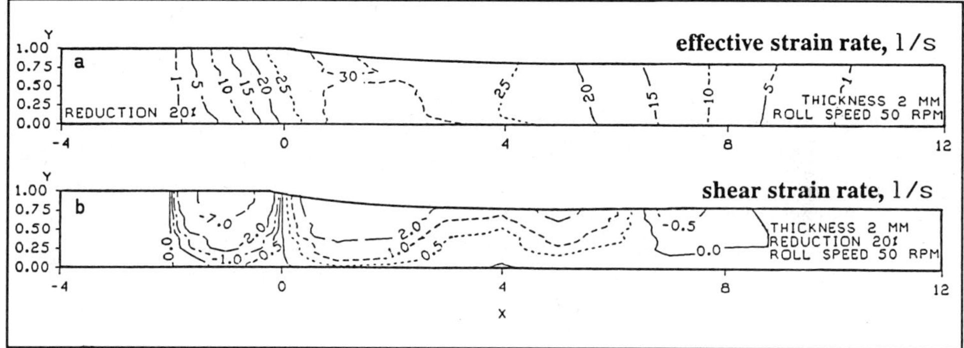

Fig.4.12 Effective strain rate (a) and shear strain rate (b) distributions during cold rolling.

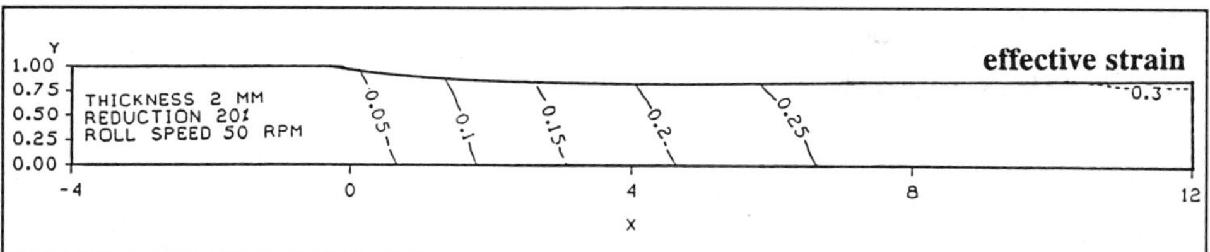

Fig.4.13 Effective strain distribution during cold rolling.

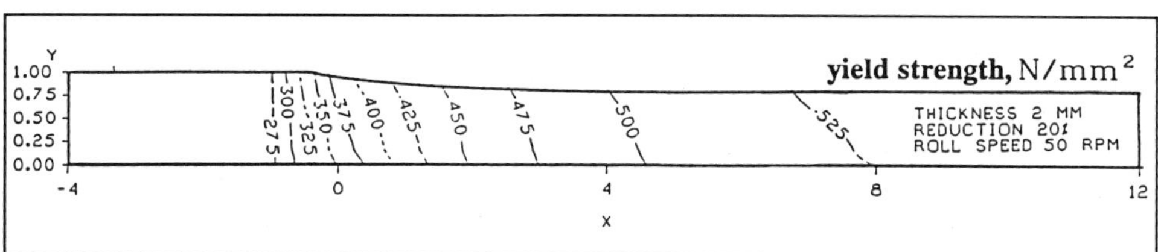

Fig.4.14 Yield strength distribution during cold rolling of a low carbon steel strip.

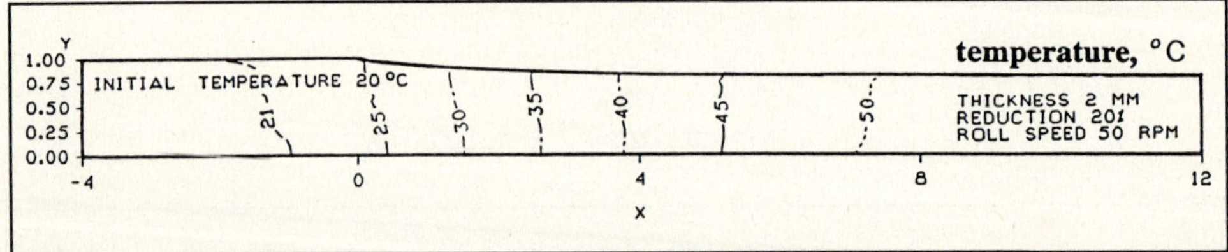

Fig.4.15 Temperature distribution during cold rolling of a low carbon steel strip.

Effective strain rate and shear strain rate fields are presented in Fig.4.12. Again, strain rate concentrations are observed close to the initial point of contact. However, the deformation is more uniform than in hot rolling (Fig.4.7). Lines of constant effective strain are shown in Fig.4.13. Due to the nonuniform distribution of strains and strain hardening the resistance to deformation varies along and across the strip - see Fig.4.14 - where the distribution of the flow strength is presented. The temperature field in the deformation zone is shown in Fig.4.15. Contrary to the hot rolling process, the warmest part of the strip is the region close to the contact surface, resulting from the enhanced effect of the heat generated by both the friction losses and redundant plastic work, near the initial point of contact. Cumulative influence of these two factors exceeds cooling by contact with rolls, which is relatively low in the cold rolling process.

4.4 Experimental Substantiation of the Model's Predictions

Experimental studies of the temperature distribution in a hot strip and/or in the work roll during rolling are not numerous. Those that have been published include the work of Sheppard and Wright (1980) who used thermocouples embedded in aluminium slabs to monitor the temperature variation during processing. Harding (1976), using several embedded thermocouples, measured the temperature history of steel slabs. Lee et al. (1963) used radiation pyrometers to measure the surface temperature of strips, processed on continuous strip mills. Steindl and Rice (1973), Jeswiet and Rice (1975) and Parker and Jeswiet (1987) used thermocouples embedded in the work roll to measure the surface temperatures in the roll gap.

As far as experimental investigation of the thermal effects in cold rolling is considered, the measurements were generally limited to the surface after rolling - see Poplawski and Seccombe (1980), Kumar and Singhal (1987) and Semiatin et al. (1987) - or to the average temperature increase during the pass, e.g. Turczyn (1981). Using a novel approach, Kannel and Dow (1974) measured surface temperature variations during the pass employing a 0.075 μm thick titanium strip deposited on the roll surface.

A broad range of hot and cold rolling experiments was conducted in the Manufacturing Processes Laboratory of the University of New Brunswick in Fredericton, Canada. Detailed description of the laboratory rolling mill and additional equipment is given in Chapter 2. The data acquisition system for the temperature measurements is described by Karagiozis (1986). Briefly, chromel-alumel (Type K) thermocouples of 1.59 mm diameter with ungrounded, exposed junctions and inconel sheaths were chosen for the temperature measurements in the hot rolling process. They were embedded to a depth of 25 mm in the tail ends of the slabs. Fig.4.16 shows the locations of the thermocouples and it is to be noted that three of them were placed as near the surface as possible. The fourth thermocouple was located at the centre-line. The distance between the surface and the centre of the thermocouple closest to it was 2.5 mm for 38 mm thick samples and 1.8 mm for 19 mm thick samples. Since some of the samples were rolled several times and the relative locations of the thermocouples changed during rolling, the actual distance between the surface and the thermocouple hole is given in the figures for each set of results. The initial diameter of the thermocouple hole was 1.6 mm in all cases.

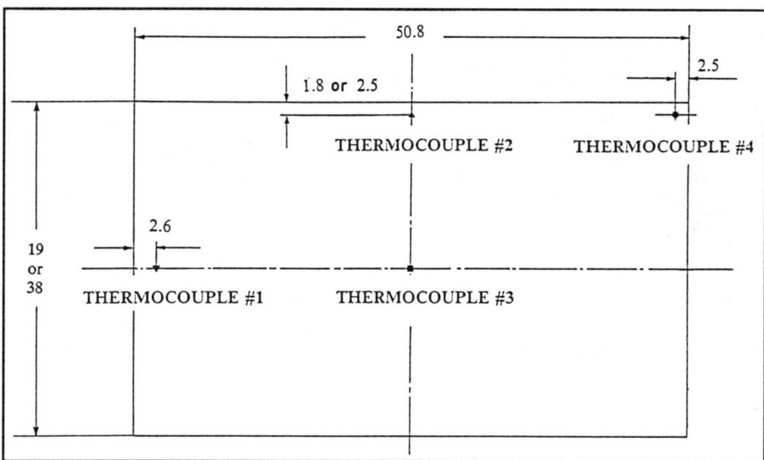

Fig.4.16 Locations of the thermocouples.

The slabs, with their thermocouples, were heated in a prewarmed laboratory furnace, whose door had a slot cut into it to allow easy insertion or removal. They were heated in air for approximately 30 minutes at which time the data acquisition system indicated that a uniform temperature distribution - to within ± 1°C - was reached. The slabs were then removed and placed in the roll gap in about 5 seconds. Temperature measurements were taken at 200 or 300 millisecond intervals and the data were stored for further processing.

Aluminum samples measuring 18.4 x 60.2 x 108 mm and 13 x 60.2 x 108 mm were used in the cold and warm rolling experiments. One type T (copper-constantan) thermocouple was embedded centrally in the tail ends of the strips during the former experiments. Two thermocouples were used in the warm rolling tests with one embedded centrally and the second located 2 mm below the surface. Independently, the roll force, roll torque, roll rpm and roll gap separation were measured and all data were monitored using the data acquisition system described in Chapter 2.

4.4.1 Hot Rolling of Steel

Typical comparisons of calculated and measured temperatures during hot rolling of low carbon steel slabs are presented in Figs 4.17 - 4.23. Fig.4.17 shows the results of calculations employing the non-steady state model, which does not account for the roll temperature rise, using $\alpha = 4800\ W/m^2K$ (Fig. 4.17a) and based on the steady-state model with convection calculating the roll temperature field, using $\alpha = 13000\ W/m^2K$ (Fig.4.17b). The non-steady state model evaluates the temperature distribution in the whole deformation zone as shown by Pietrzyk and Lenard (1988) and, in consequence, a comparison for three thermocouples is possible. Somewhat better agreement between measured and calculated cooling rates is obtained using the steady state version of the model. This version is used in all further calculations.

Measurements for three passes of 38 mm thick slabs are compared with predictive calculations, with $\alpha = 13000\ W/m^2K$, in Figs 4.18,4.19 and 4.20. In Figs 4.17 and 4.18 approximately 20% reduction was obtained at average entry temperatures of 850°C and 760°C, respectively. The speed of rolling was 4 rpm. In both comparisons the differences between measurements and predictions are less than 5%, indicating that the predictive ability of the model is reasonable. The predicted rates of cooling and heating are also in good agreement with the observations.

The model also calculates the temperature rise due to plastic work reasonably well. After reheating, some of the samples were rolled again with lighter reductions. The results of experiment with 10% and 8% reductions respectively, are presented in Figs 4.19 and 4.20 and again adequate accuracy of the predictions is evident. The predicted and observed rates of heating and cooling are somewhat less close than those for the heavier reductions.

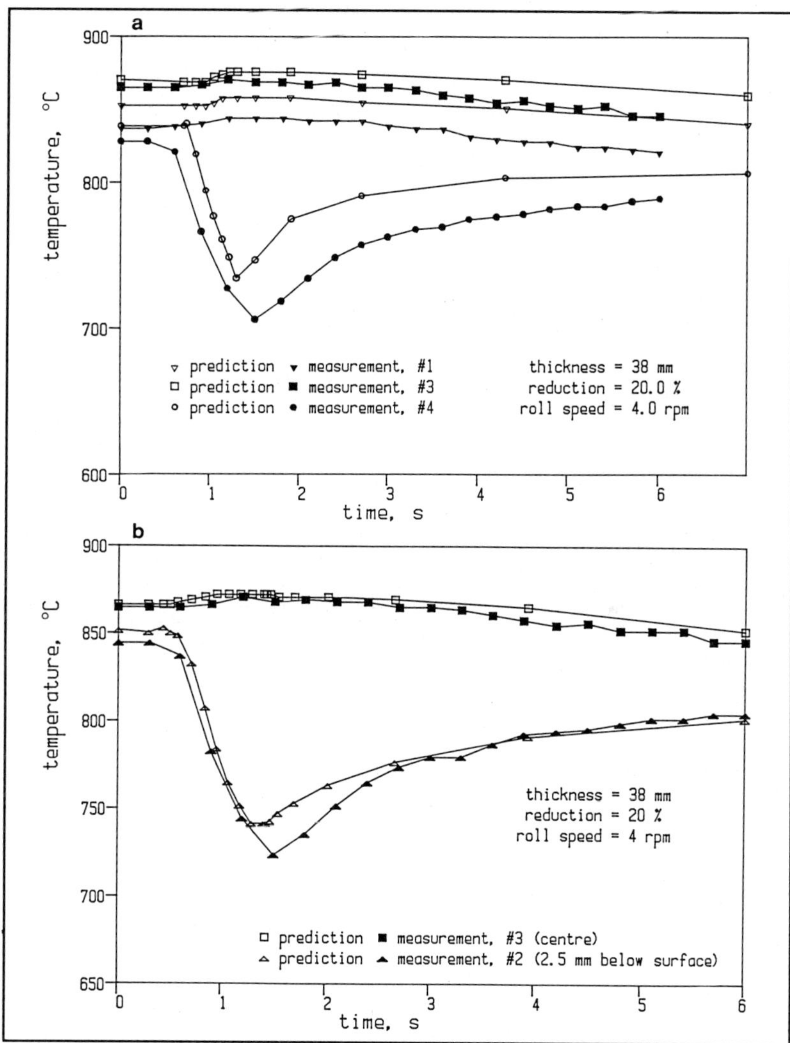

Fig.4.17 Experimental substantiation of time-temperature profiles calculated using the non-steady state model with constant roll temperature (a) and the steady state model which accounts for the roll temperature build up (b).

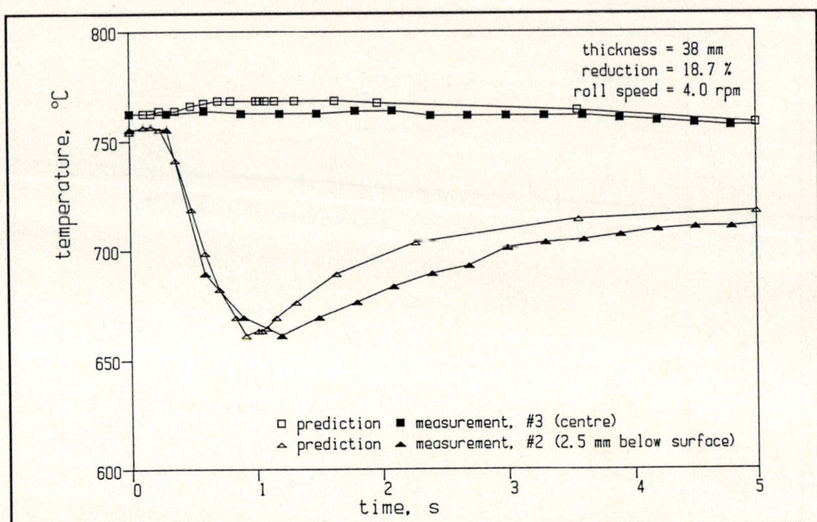

Fig.4.18 Time-temperature profiles for a 38 mm thick slab reduced by 18.7%.

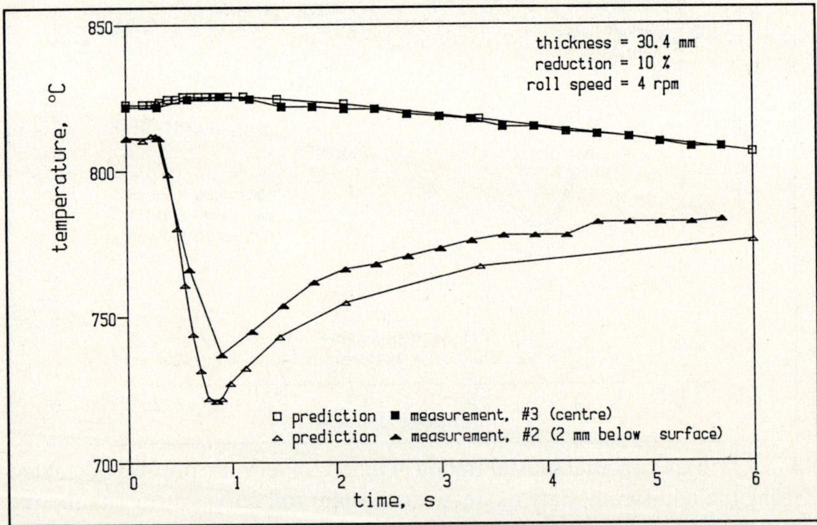

Fig.4.19 Time-temperature profiles for a 30.4 mm thick slab reduced by 10%.

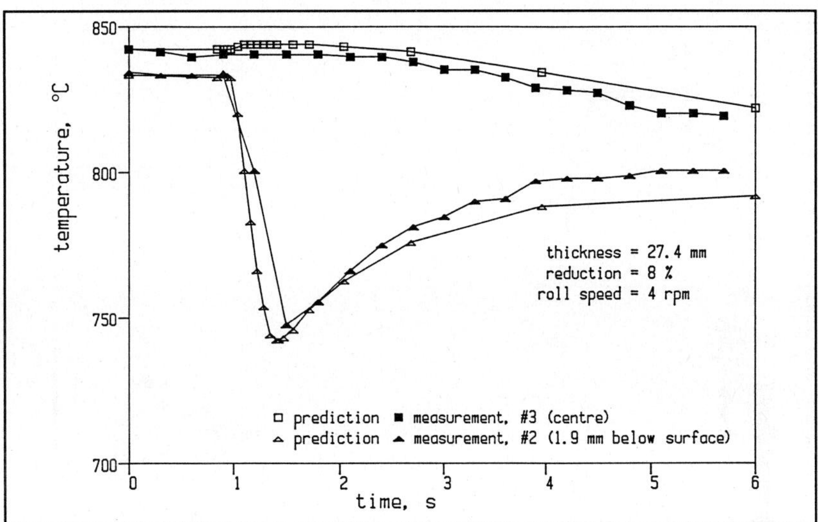

Fig.4.20 Time-temperature profiles for a 27.4 mm thick slab reduced by 8%.

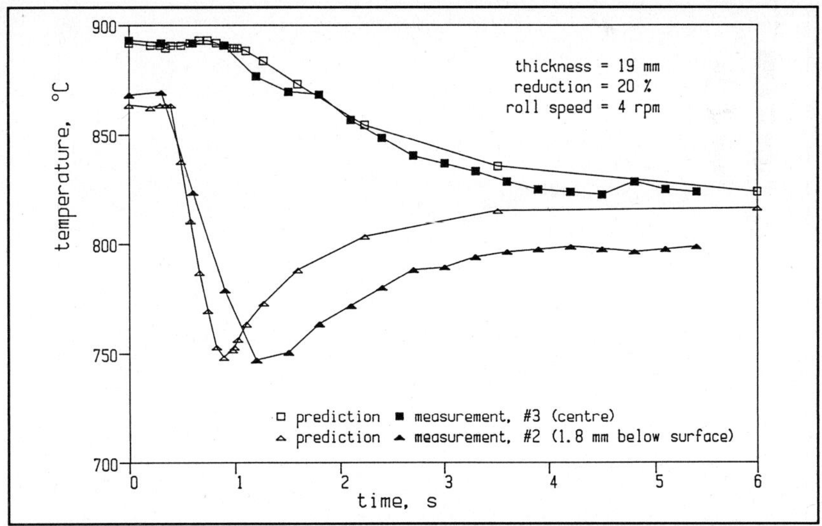

Fig.4.21 Time-temperature profiles for a 19 mm thick slab reduced by 20%.

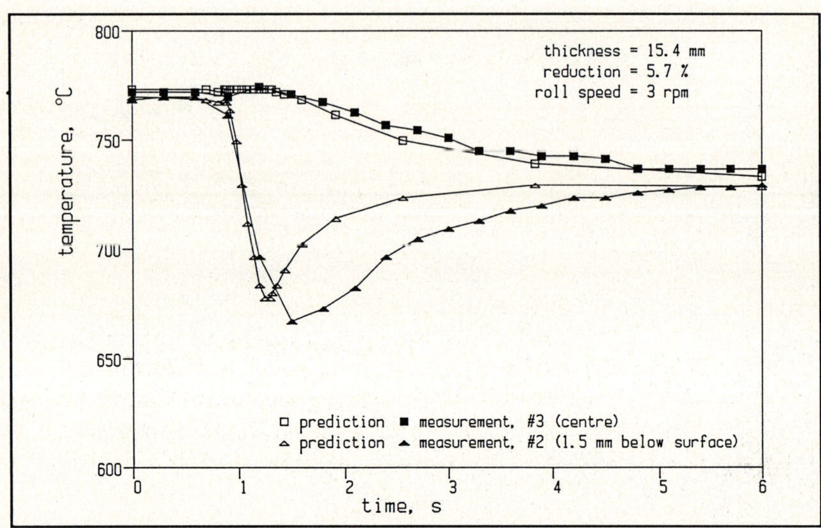

Fig.4.22 Time-temperature profiles for a 15.4 mm thick slab reduced by 5.7%.

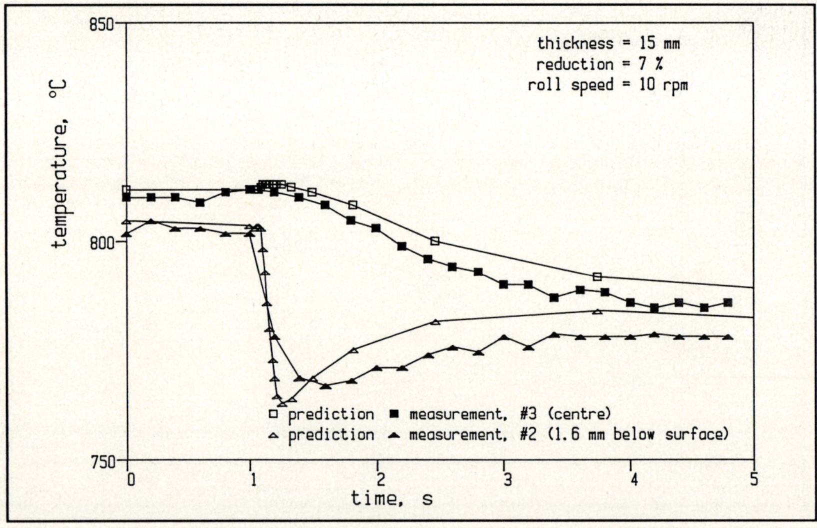

Fig.4.23 Time-temperature profiles for a 15 mm thick slab reduced by 7%.

Similar comparisons of the measured and predicted temperatures for the 19 mm, 15.4 mm and 15 mm thick slabs are presented in Figs 4.21,4.22 and 4.23, and the conclusions are similar to those mentioned above. The differences between the predicted and measured cooling rates are somewhat larger than for the 38 mm slabs and it is believed that the relative sizes of the thermocouples and the strip thicknesses are responsible, especially near the surface where cooling rates and temperature gradients are fairly high. While no surface temperatures have been recórded during the test, some feel for the near-surface temperature gradients may be arrived at by considering results of computations given in Figs 4.11 and 4.24. Fig.4.11 shows the temperature distribution in the work roll and in the strip during 20% reduction of a 19 mm thick sample. Sharp temperature gradients are observed close to the contact surface. Temperature variations across the strip thickness at the exit plane are compared with that obtained for the 38 mm thick sample, reduced by 18.7%, in Fig.4.24. The locations and dimensions of the thermocouple holes are also shown there. The thermocouple in the thinner sample is located closer to the surface and, due to the higher temperature gradient there, the temperature difference across the diameter of the hole is about 107°C. For the 38 mm thick sample approximately 70°C is calculated. Further substantiation of the model is presented by Pietrzyk and Lenard (1988, 1989a, 1989b) and by Lenard and Pietrzyk (1989).

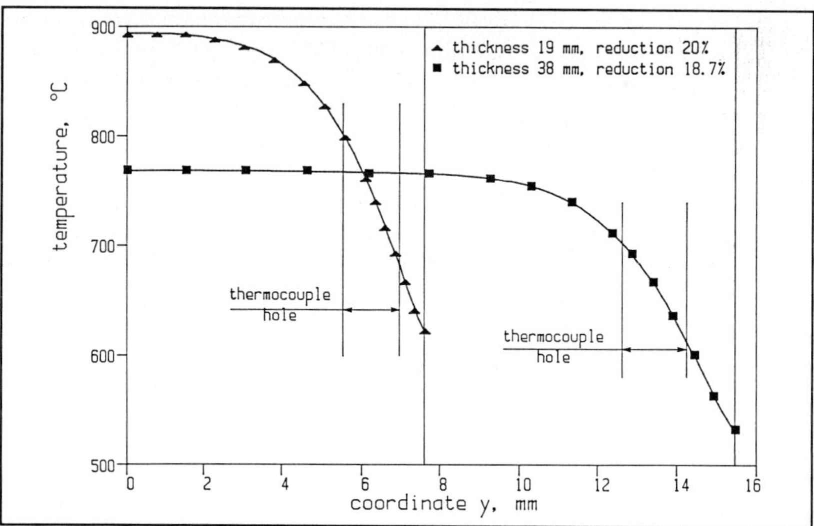

Fig.4.24 Temperature distribution across the sample at the exit plane and thermocouple locations for rolling of 19 and 38 mm thick slabs.

Comparison of the predicted values of the temperature field and the experimental results highlighted some important considerations of the testing technique. These include the knowledge of the locations of the thermocouples before and after reduction; the need for sufficiently low sampling times during measurements and the use of driving motors of sufficient power. A thorough discussion of these is presented by Pietrzyk and Lenard (1988) and by Pietrzyk et al. (1989).

Predictions of the thermal-mechanical model were also tested against experimental data obtained during continuous hot rolling of strips in industrial conditions. Rolling parameters, typical results of calculations and measurements for a seven-stand continuous mill are given in Table 4.2. In this table h represents the current thickness of the strip, r is the reduction, T is the strip temperature at the exit, n is the rotational speed of the roll, R is the roll radius, R' is the radius of the deformed roll, l_d is the length of contact for rigid rolls, l'_d is the length of contact with the flattened roll, s is the forward slip, v_r is the roll velocity and v_2 stands for the strip exit velocity. In the two columns listing the average temperatures the first one, marked "Ind", gives the industrial data. The second column contains the numbers computed by the present technique. Both begin with 1058°C, the temperature measured on the transfer table. In the calculations cooling in air for 30 seconds is then simulated, resulting in 988°C at the exit from the first pass, close to the temperature of the industrial data. Part of the success of the present model is determined by its ability to predict the temperature at the exit from the seventh stand, measured to be 889°C. In the calculations a heat transfer coefficient $\alpha = 50000$ W/m^2K is used, chosen following Murata et al. (1984) and Pietrzyk and Lenard (1989b). Forward slip, also given in Table 4.2, is calculated using Fink's formula (Pavlov, 1955 and Morawiecki et al., 1986).

All results, given in Table 4.2, are obtained for a low carbon steel containing 0.07%C, 1.26%Mn, 0.421%Si, 0.013%Cu, 0.03%Ni, 0.03%Cr and 0.075%V. In the calculations of the flow strength (Shida, 1974) the carbon equivalent was taken to be 0.3 (DeArdo, 1988). As observed, predictions based on the slab method and on the finite element approach are close to each other. Comparison of measurements and calculations of roll separating forces show reasonably good agreement for passes 1,3,4 and 5. Analysis of results for pass 2 indicates difficulties with the data since the lower reduction in this pass in comparison with pass 1 did not result in comparable decreases of the measured roll force and roll torque. In the last two passes measured roll forces and torques are significantly larger than the predicted ones. The model which assumes full recrystallization in each pass is responsible. Access to reliable mathematical accounts of the evolution of the microstructure during finish rolling should decrease the differences.

Material	Process	Yield stress $\frac{N}{mm^2}$	Conductivity $\frac{W}{mK}$	Density $\frac{kg}{m^3}$	Specific heat $\frac{J}{kgK}$
Low carbon steel	hot	Shida's (1974) eq.	$23.16+51.96\exp(-2.025\overline{T})$	$\frac{7850}{(1+0.004\overline{T}^2)^3}$	$689.2+46.16\exp(3.78\overline{T})$
	cold	$265(1+250\epsilon)^{0.162}$			$207.9+294.4\exp(1.412\overline{T})$
Stainless steel 17.5%Cr 8.85% Ni	hot	Shida's (1974) eq.	$21.85+14.32\overline{T}^{1.831}$	$\frac{7850}{(1+\gamma_1)^3}$	$474.4+196.7\overline{T}+3.786\overline{T}^2-59.56\overline{T}^3$
	cold	$889.6(\epsilon+0.03)^{0.286}\exp\frac{-0.007}{\epsilon+0.03}$			
Pure aluminium	cold	$44(1+87\epsilon)^{0.2605}$	$33\overline{T}+236$	$\frac{2840}{(1+\gamma_2)^3}$	$836.8+620\overline{T}$
Aluminium alloy	cold	$229(1+308\epsilon)^{0.1222}$	$33\overline{T}+236$	$\frac{2840}{(1+\gamma_2)^3}$	$836.8+620\overline{T}$

$\overline{T}=\frac{T}{1000}$ $\quad \gamma_1=-0.334+0.836\overline{T}+1.169\overline{T}^2+0.00827\overline{T}^3$ $\quad \gamma_2=-0.000457+0.02343\overline{T}+0.003493\overline{T}^2+0.00827\overline{T}^3$

Table 4.1 Functions describing properties of various materials.

pass no	h	r	T °C		n	R	R'	l_d	l_d	s	v_r	v_2	force kN/mm			torque Nm/mm			t
	mm		Ind.	FEM	rpm	mm	mm	mm	mm		m/s	m/s	slab	FEM	meas	slab	FEM	meas	s
0	38.8	-	1058	1058	-	-	-	-	-	-	-	-	-	-	-	-	-	-	-
1	21.6	0.44	993	988	32	360	389	78.7	81.8	0.178	1.21	1.42	14.5	16.4	14.3	505	636	400	3.85
2	14.4	0.33	986	973	52	336	379	49.2	52.2	0.116	1.83	2.05	10.6	12.3	17.1	228	288	414	2.67
3	8.6	0.40	968	957	80	353	398	45.3	48.1	0.158	2.96	3.43	13.3	14.6	13.4	257	304	226	1.60
4	6.1	0.29	951	938	116	343	409	29.3	32.0	0.098	4.16	4.57	9.7	10.5	12.8	116	137	160	1.20
5	4.7	0.23	935	922	155	388	494	23.3	26.3	0.072	6.30	6.75	8.9	9.4	11.6	80	95	104	0.81
6	3.9	0.17	913	904	184	348	456	16.7	19.2	0.050	6.71	7.04	6.2	6.4	9.2	42	46	60	0.78
7	3.5	0.10	898	894	204	369	413	13.6	14.4	0.028	7.87	8.10	4.0	3.7	7.2	19	21	70	-

Table 4.2 Measured and calculated rolling parameters for the 7-stand production type continuous hot strip mill.

Typical results of calculations of the time-temperature profiles during continuous hot rolling of a steel strip are presented in Fig. 4.25. Temperatures in the centre, at the surface as well as the average temperature are given.

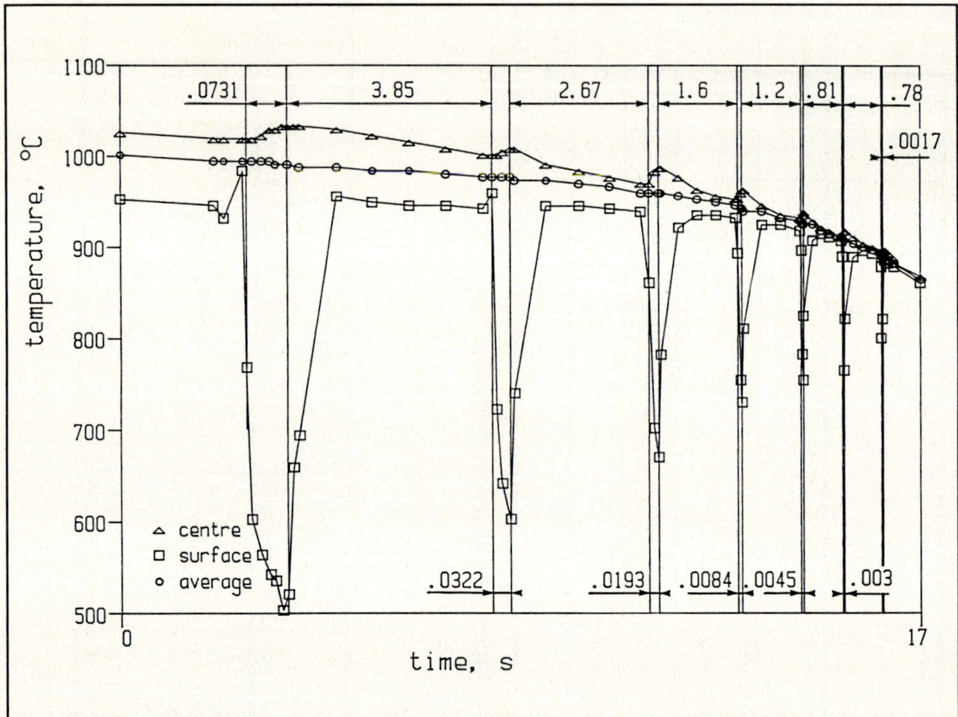

Fig.4.25 Time-temperature profiles for continuous hot rolling of a strip.

4.4.2 Warm Rolling of Aluminium

The steady state model was also validated by comparing its predictions to measurements obtained during warm rolling of aluminium samples in laboratory conditions, using the experimental equipment described in Chapter 2. Two type K thermocouples were embedded in the tail end of each sample, one in the centre and one located 2 mm below the surface, near a corner and, as before, the temperature changes during the tests were monitored by a microprocessor based data acquisition system. Typical results of the comparison between measured and calculated time-temperature profiles are presented in Figs 4.26 and 4.27. A heat transfer coefficient $\alpha = 30000\ W/m^2K$, established by Pietrzyk and Lenard (1989b), is used in these calculations. Reasonably good agreement between measured and calculated cooling rates is observed in both figures. There is little indication of deformation heating in these cases, no doubt due to the low resistance to deformation of the commercially pure aluminium. The temperature increase of the roll seems to be of some importance in the warm rolling process, however. Theoretical time-temperature profiles for the roll, presented in Fig.4.28, show a 100°C increase of the surface temperature as well as very high heating rates, indicating a significant and fast decrease of the temperature difference between the surface of the roll and the sample. This is, of course, reflected in the value of the heat transfer coefficient which is used in the calculations, as shown in the next section and by Pietrzyk and Lenard (1989c).

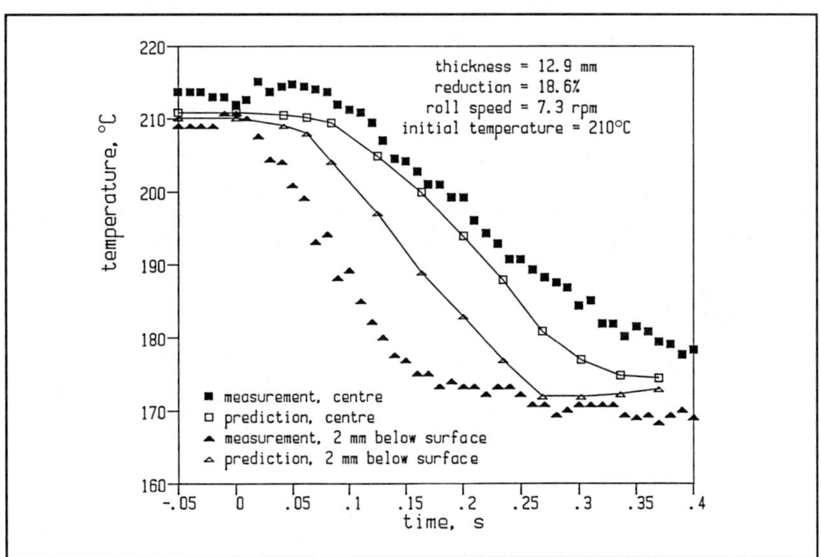

Fig.4.26 Time-temperature profiles for warm rolling of aluminium at 210°C.

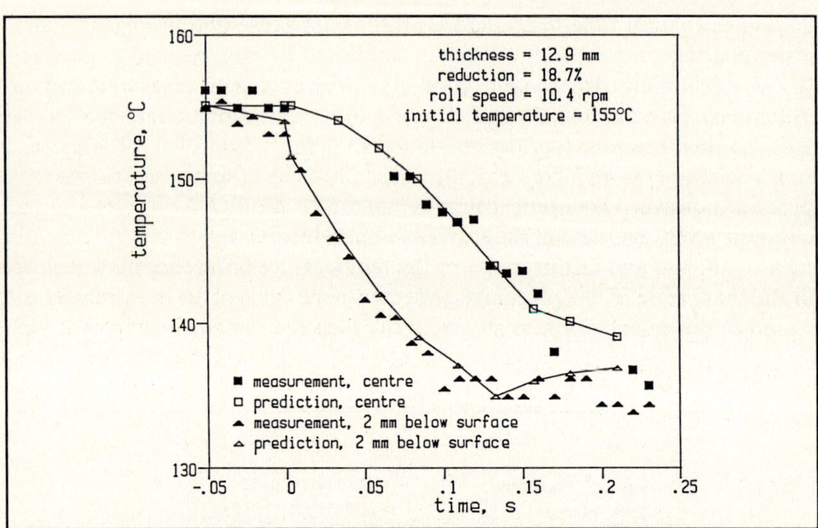

Fig.4.27 Time-temperature profiles for warm rolling of aluminium at 155°C.

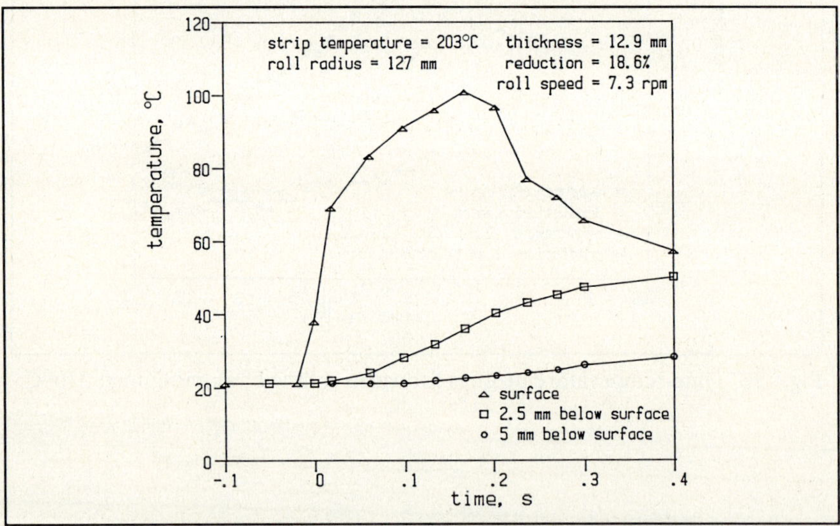

Fig.4.28 Calculated roll temperature increase during warm rolling of aluminium.

4.4.3 Cold Rolling of Aluminium

Further comparisons of calculations and measurements during unlubricated cold rolling of aluminium samples are shown in Figs.4.29 and 4.30. All materials properties, necessary for heat transfer calculations, are given as functions of the temperature in Table 4.1. The friction coefficient μ is assumed to be 0.12.

The first set of calculations and measurements is carried out for a reasonably tough alloy, the strain hardening curve for which is given in Table 4.1. The temperature rise at the centre of the strip, during 5.5% reduction, is shown in Fig.4.29. The predictions of the mathematical model are also presented, the computations being carried out by both the non-steady state model, given by equation (4.35), and by the steady state model with the convective term, represented by equation (4.39). It is observed that the difference between the measured and computed temperature values at exit is approximately 1.5^{o}C. There appears to be a time delay of 0.4 seconds between the thermocouple's response and the rate of temperature rise as predicted by the model. This is most likely caused by the loss of contact and the increased contact thermal resistance between the tip of the thermocouple and the bottom of the hole in which it is embedded. Care was taken that the bead and the strip be in contact on insertion. During rolling however, the softer strip and thus the thermocouple's hole would elongate more than the thermocouple itself and an air gap, acting as an insulator, is created, contributing to the loss of contact.

The already reduced strip was rolled once again, reducing it by a further 10.2%. The predicted and measured temperature-time profiles during that pass are given in Fig.4.30. The difference between the computed and measured temperature values is of the same order of magnitude as in the first pass. The time delay of the thermocouple has decreased substantially, indicating a possible reduction of the air gap and an attendant growth of the contact surface between the thermocouple's tip and the rolled metal.

It is observed from Figs 4.29 and 4.30 that the steady state approach with the convective term gives the temperature vs. time curves somewhat closer to the experimental ones than the non-steady state model, especially in the second pass where probably much better bead to strip contact was maintained. It may be concluded then that in the cold rolling process the longitudinal heat conduction is of significant importance (Pietrzyk and Lenard, 1989a). In all further examples of calculations, presented in this section, the steady-state convective-diffusion model is used. It is very likely that the reasonably high resistance to deformation and the attendant low ductility of the aluminium alloy used in the first set of experiments was responsible for the pronounced loss of contact between the thermocouple and the material

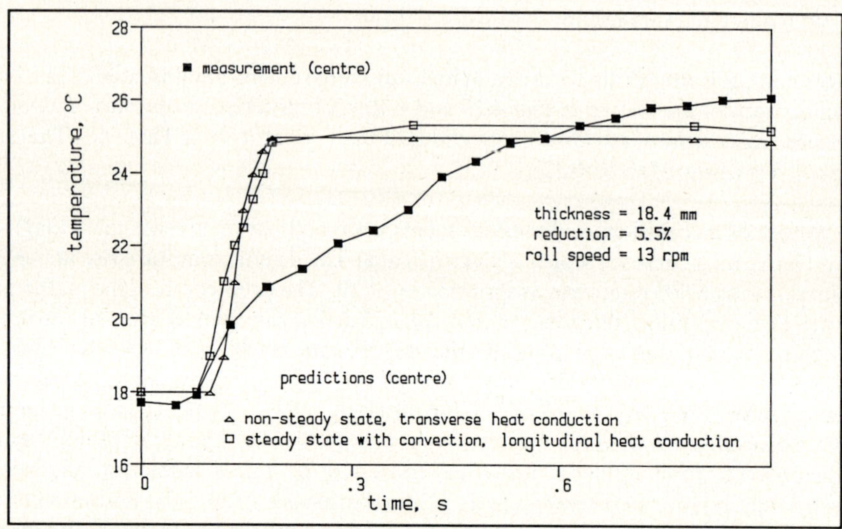

Fig.4.29 Temperature rise during cold rolling of an aluminium alloy, pass #1.

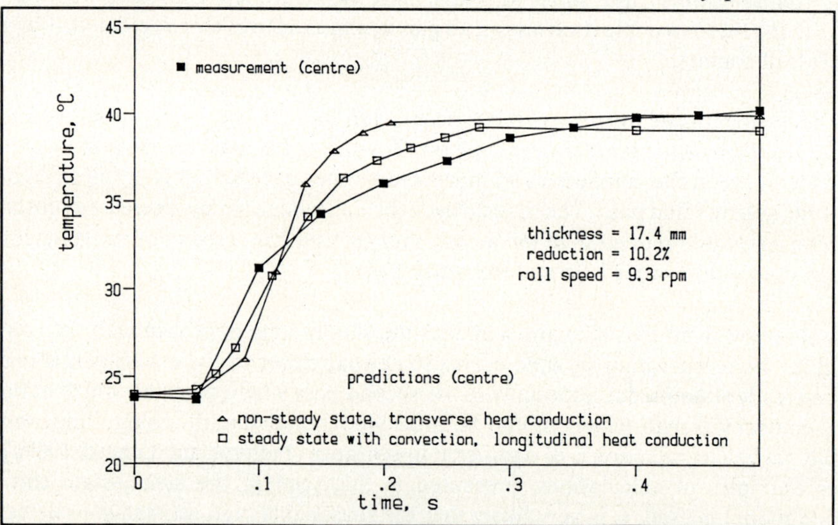

Fig.4.30 Temperature rise during cold rolling of an aluminium alloy, pass #2.

and, in consequence, for the time delay between the thermocouple response and the predictions. In order to avoid this problem a commercially pure, annealed aluminium was used in all further experiments. The strain hardening curve of the aluminium, determined in compression tests, is given in Table 4.1 on page 132. The low resistance to deformation allowed much larger reductions per pass and total reductions. Typical results of measurements and calculations of temperature using the softer aluminium are shown in Figs 4.31 and 4.32. As well, in Fig. 4.32 the results are compared with the final temperatures, calculated using the closed form formula of Roberts (1978)

$$\Delta T = \frac{\left(1-\frac{r}{4}\right)\sigma_p \ln\frac{r}{1-r}}{\left(1-\frac{r}{2}\right)\rho c_p} \tag{4.42}$$

where ΔT is the temperature increase, σ_p is the yield strength, c_p is the specific heat, ρ is the density and r represents the reduction per pass.

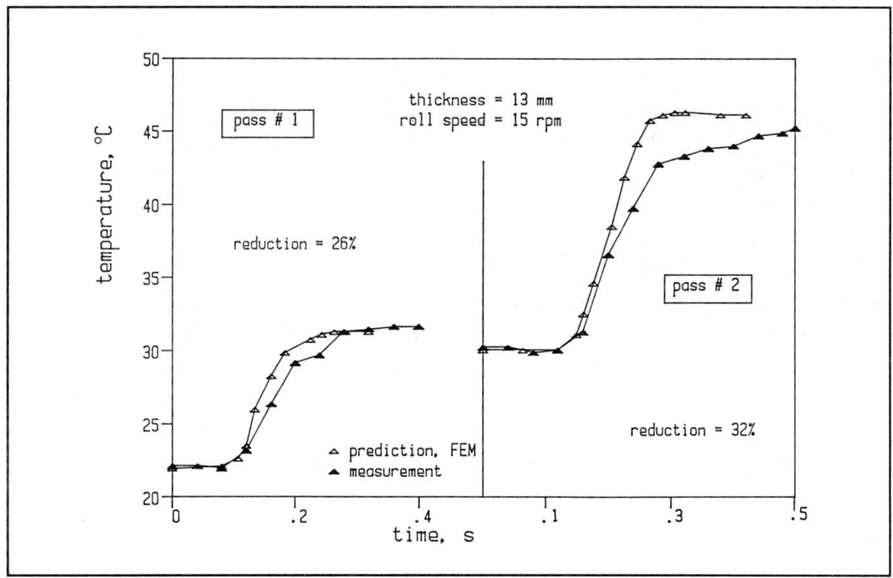

Fig.4:31 Temperature rise during cold rolling of commercially pure aluminium.

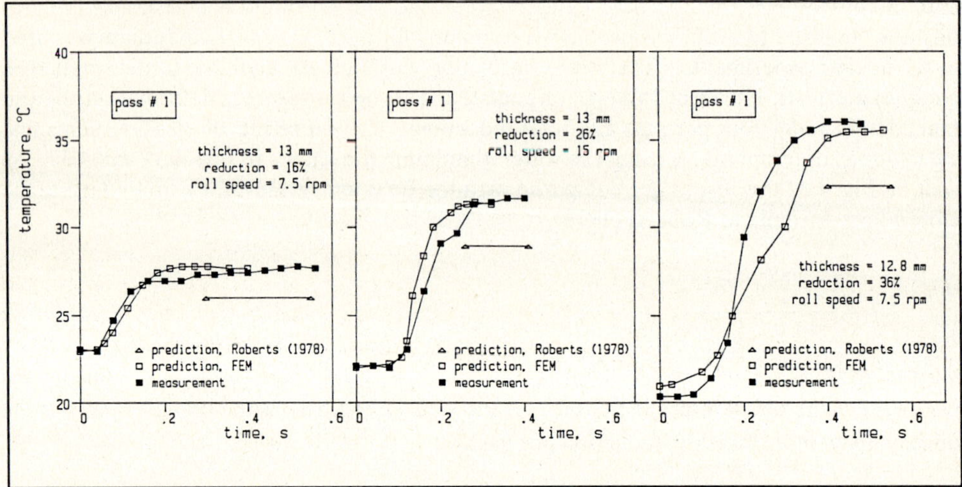

Fig.4.32 The effect of reduction on deformation heating during cold rolling of commercially pure aluminium.

Fig.4.31 shows time-temperature profiles during cold rolling of a 13 mm thick slab in two subsequent passes. Very good agreement between experimental and theoretical curves is observed. The lower rate of temperature rise in the final stages of the second pass is again probably caused by the partial loss of contact between the thermocouple bead and the sample. The effect of the reduction on the temperature rise can be observed in Fig. 4.32. Roberts' (1978) formula (equation 4.42) gives temperature increases lower than the measurements or the predictions in all cases. That could be expected, because (4.42) was derived using an assumption of uniform deformation and redundant strains were neglected. Analysis of results presented in Figs 4.29 - 4.32 leads to several observations about deformation heating during the cold rolling process. Comparison of Fig.4.32c and the second pass in Fig.4.31 shows the effect of strain hardening on the heat generated during plastic deformation, with 36% reduction in the first pass after annealing (Fig.4.32c) giving a lower temperature increase than the 32% reduction in the second pass (Fig.4.31). The effect of the material's flow strength on deformation heating can also be seen by a comparison of Figs. 4.29 and 4.30 with Figs 4.31 and 4.32. For example, 10% reduction of the tough aluminium alloy (Fig.4.30) results in a temperature rise similar to that experienced during 32% reduction of the softer metal (Fig.4.32b).

4.5 Role of the Heat Transfer Coefficient

The accuracy of predictions of mathematical models of hot, warm and cold rolling, in which thermal events are coupled with mechanical phenomena, depends on the rigour and quality of the descriptions of the boundary conditions. For the mechanical component these involve the distribution of surface tractions on the roll-strip interface and the shape of that interface. In the thermal portion the heat transfer coefficient in the roll gap needs to be specified.

As Devadas and Samarasekera (1986) write, experimentally determined values of that coefficient have not been widely reported in the literature. Those that have been published - see, for example Murata et al. (1984), Stevens et al. (1971), Harding (1976) and Preisendanz (1967) - provide numbers that differ broadly for conditions which are not too different from each other. Typical values of the heat transfer coefficient measured by Murata et al. (1984) are given in Table 4.3. However, since the number of parameters that influence the magnitude of the heat flux at the contact surface is large, choice of the heat transfer coefficient for use in modelling is still problematic.

	Heat transfer coefficient, W/m^2K	
Condition	no scale	scale 10 μm
no lubricant	29.1 - 34.9	7.0 - 10.6
water	23.3 - 81.4	10.6
hot rolling oil	200 - 460	5.8
hot rolling oil + 20%$CaCO_3$	69.8 - 175	12.8 - 23.3
hot rolling oil + 40%$CaCO_3$	12.79 - 17.4	-
KPO_3	5.8	-

Table 4.3 Heat transfer coefficients at the contact surface for various conditions, as measured by Murata et al. (1984).

The value of the heat transfer coefficient determined from the experimental data of Karagiozis and Lenard (1988) during hot rolling of steel is 4800 W/m^2K, obtained using a constant roll temperature of 40°C in the calculations. In order to assess the effect of the heat transfer coefficient on the accuracy of the model, calculations for three various values were carried out. Increase of the roll temperature was not included in these calculations. The results are compared with the measurements of the temperature variations in Fig.4.33. It is observed that the value of the heat transfer coefficient has some influence on the temperature distribution inside the deformation zone. Increasing the magnitude of α to 7000 W/m^2K overestimates the cooling rate of the strip while $\alpha = 3000$ W/m^2K results in slower cooling in comparison with the experimental data. However, shortly after the pass the temperature is distributed more uniformly and the results of calculations for various heat transfer coefficients do not differ significantly. This situation is due to a very short time of contact between the roll and the strip, which was 0.6 sec in the case considered.

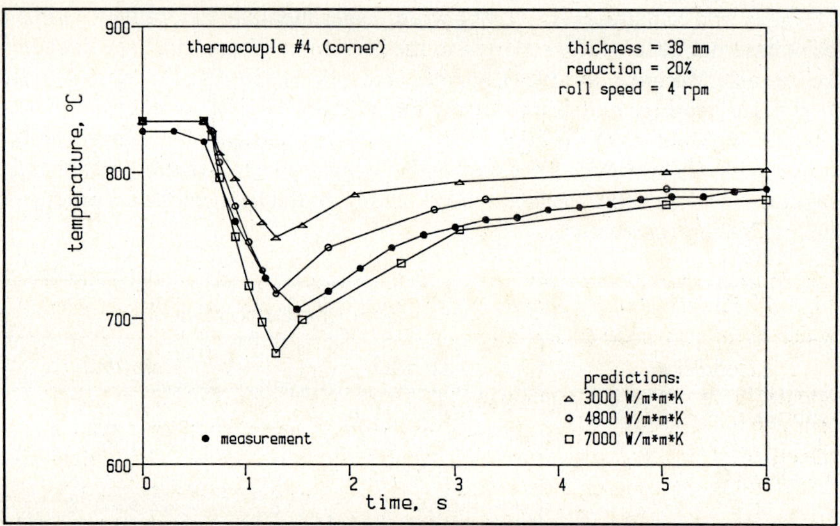

Fig.4.33 The effect of the heat transfer coefficient on the model's prediction (constant roll temperature).

In the hot rolling process the roll temperature increases rapidly and that increase has a significant influence on the heat transfer between the strip and the roll. As shown in Fig.4.34, calculations based on the model which includes evaluation of the increase of the roll's temperature underestimate the cooling rates for the heat transfer coefficient $\alpha = 4800$ W/m^2K and the best agreement between measured and calculated time-temperature profiles is obtained for $\alpha = 13000$ W/m^2K. This value is therefore suggested for the simulation of hot rolling of steel in laboratory conditions, when models which account for the temperature rise of the roll are used.

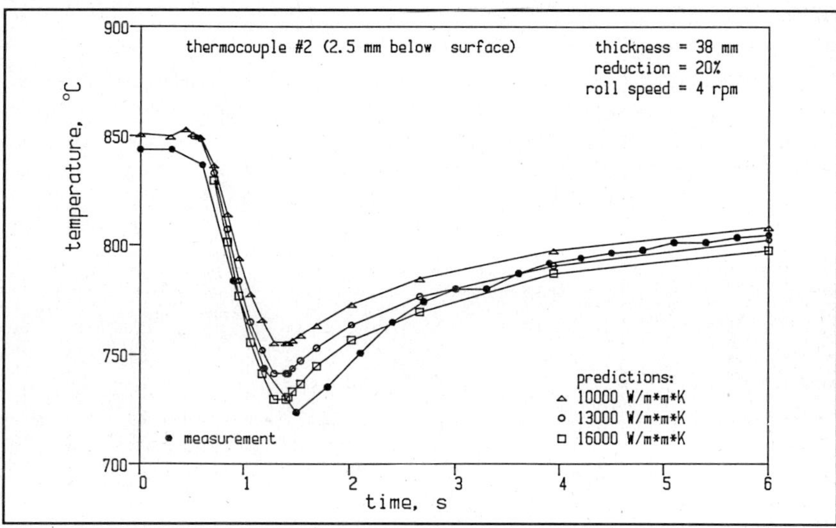

Fig.4.34 The effect of the heat transfer coefficient on the predictions of the model which accounts for the temperature increase of the roll.

In hot strip mills the effect of changes in the heat transfer coefficient on the temperature variations can become cumulative and significant differences of predicted temperatures may be obtained after the last stand. To test this hypothesis, predictions of the model were tested against temperature measurements during hot rolling of a strip in a seven stand hot strip mill. Calculations were carried out for three different heat transfer coefficients. (Fig.4.35). It is noted that the heat transfer coefficient $\alpha = 4800\ \mathrm{W/m^2K}$, used successfully in laboratory conditions, underestimates the cooling rates significantly. The value of $\alpha = 50000\ \mathrm{W/m^2K}$ gives a reasonably good prediction of the final strip temperature after the 7th stand, which was measured to be 889°C.

Further analysis of the effect of the heat transfer coefficient on the predictions of temperatures includes the process of warm rolling of aluminium samples. Average heat transfer coefficients, calculated by Pietrzyk and Lenard (1989b) from the results of temperature measurements, varied between 18500 and 21500 $\mathrm{W/m^2K}$. That is somewhat lower than the value of 30000 $\mathrm{W/m^2K}$ reported by Smelser and Thompson (1987) for rolling of aluminium slabs at 250°C but close to the values of 15000 - 20000 $\mathrm{W/m^2K}$, measured by Semiatin et al. (1987) in a ring compression test. The effect of the heat transfer coefficient on the predictions

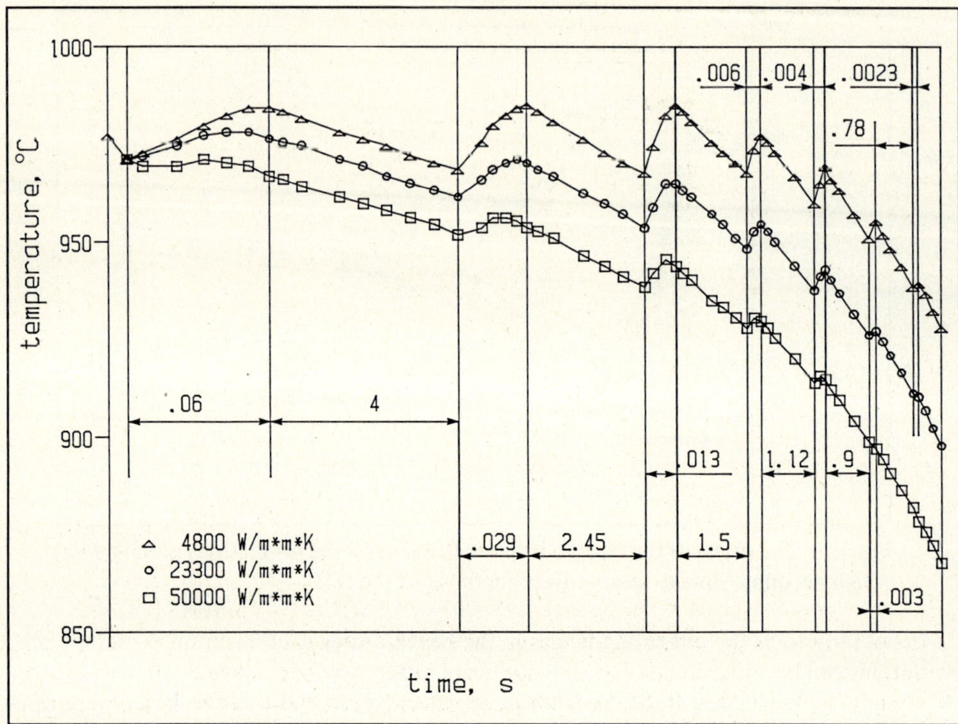

Fig.4.35 The effect of the heat transfer coefficient on the temperature calculations during continuous hot rolling of strips.

of the finite element model, which does not account for the temperature rise of the roll, is shown in Fig.4.36, where calculations for $\alpha = 4800$, 20000 and 30000 W/m^2K are compared with measurements carried out during warm rolling of a 12.9 mm thick sample at an initial temperature of 210°C, reduced by 18.6% at a roll speed of 7.3 rpm. It is observed that the assumption of constant roll temperatures leads to very good agreement between measured and calculated cooling rates for $\alpha = 20000$ W/m^2K. A heat transfer coefficient of 4800 W/m^2K, appropriate for hot rolling of steel in laboratory conditions, underestimates the cooling rates during warm rolling, while $\alpha = 30000$ W/m^2K results in faster cooling. Similar results based on the model which accounts for the roll temperature rise are presented in Fig.4.37, showing that the best agreement between measured and calculated time-temperature profiles is obtained for $\alpha = 30000$ W/m^2K.

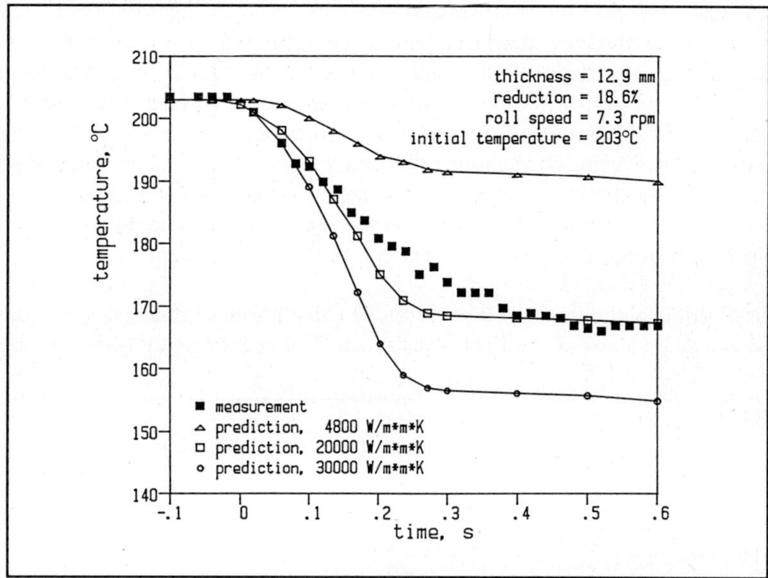

Fig.4.36 The effect of the heat transfer coefficient on the model's predictions for constant roll temperature in warm rolling.

Due to a smaller temperature difference between the strip and the roll, the effect of the heat transfer coefficient in cold rolling processes is of less importance than in warm and hot rolling. Nevertheless, as the temperature of the workpiece increases, cooling by the roll becomes more intense and that makes the choice of the heat transfer coefficient for use in modelling of some importance. The temperature rise of the strip, measured in cold rolling, varied between a few degrees C and about 30°C. However, due to the small effect of cooling by contact with the roll it is difficult to calculate the heat transfer coefficient directly from the temperature measurements. Calculations of the temperature variations during cold rolling of a 13 mm thick aluminium sample (Fig.4.38) show a very small effect of the heat transfer coefficient on the results. It is concluded then, that in the cold rolling process, the heat transfer coefficient determined either during warm rolling or by techniques such as that of Semiatin et al. (1987) or Murata et al. (1984) could be used. Semiatin et al. (1987) developed experimental and analytical techniques for the determination of this coefficient in nonisothermal forming processes. They presented results for temperatures ranging from 400°C to 800°C. Some of these results, obtained from a two-die experiment, are shown in Table 4.4. Semiatin et al. (1987) also used the ring compression test to evaluate the heat transfer coefficient. On comparing the temperatures measured during compression of an aluminium ring to the calibration curves, they deduced the heat transfer coefficient to be approximately 7500 W/m^2K and 17000 W/m^2K for the slow and fast tests, respectively. These values are suggested for use in simulation of thermal events during cold rolling of both aluminium and steel in a single pass. During continuous cold rolling process the effect of deformation heating is cumulative and the strip's temperature may rise by as much as several hundreds degrees C - see Roberts (1978), Pietrzyk et al. (1982) and Turczyn and Nowakowski (1985). In consequence, the heat transferred to the roll becomes more pronounced and hence, more important.

Analysis of published data as well as results of calculations and measurements lead to the conclusion that the values of the heat transfer coefficient published in the literature should

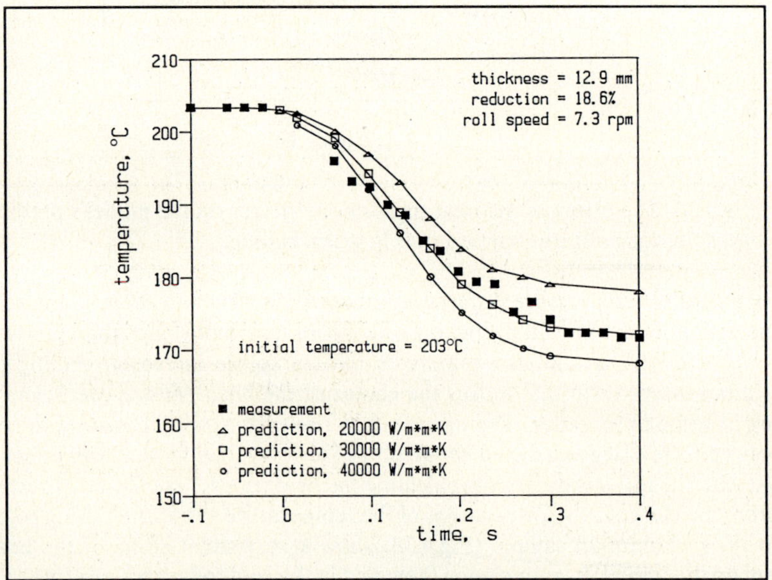

Fig.4.37 The effect of the heat transfer coefficient on the predictions of the model which accounts for the roll temperature increase in warm rolling.

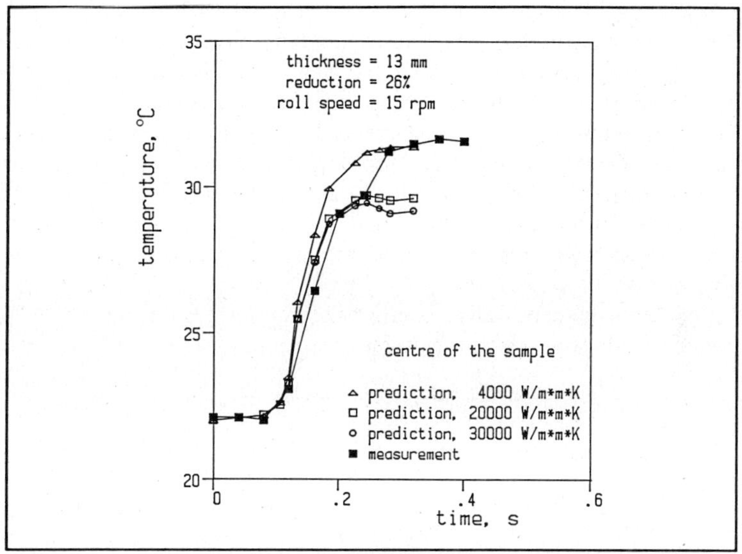

Fig.4.38 The effect of the heat transfer coefficient on the temperature calculations in the cold rolling process.

lubricant	pressure, N/mm^2					
	0	0.03	0.85	14	85	150
none	400	750	1500	4000	7500	7500
renite S28	500	900	4000	6500	7500	7500
wynn 880n	400	1900	4000	7000	7500	7500

Table 4.4 Heat transfer coefficients (α, W/m^2K) measured by Semiatin et al. (1987) in two-die experiments for various lubricants and die pressures.

not be employed directly in the mathematical models. The choice of that coefficient should be made considering the model and the boundary conditions in addition to the relevant process parameters.

CHAPTER 5

THE ROLE OF THE SHAPE COEFFICIENT IN MODELLING OF THE FLAT ROLLING PROCESS

A common feature of some metal forming operations, such as rolling, drawing or extrusion, is the flow of metal through a converging channel. Changes in the displacement and velocity components are unavoidable in the material travelling through this channel. These changes also tend not to be the same for all particles of the material. When they differ, as they generally do, gradients of strain and strain rate are the result.

A basic geometric feature of each channel is the mean thickness-to-length ratio of the plastic zone that fills it. This ratio is defined by Backofen (1972) as the shape coefficient of the deformation zone, Δ. It was shown by several authors - see Tselikov and Grishkov (1970), Backofen (1972), Wright (1976), Blazynski (1979), Pietrzyk (1983b, 1986), Pietrzyk and Kusiak (1986) and Skolyszewski (1986) - that the flow of metal in various processes depends strongly on the shape coefficient.

The effects of Δ on the tool pressure have been widely investigated. One trend, which has already become apparent, is that in frictionless processes the pressure normal to the tool-material interface is predicted to increase with Δ. The theoretical solution of this problem for the full range of Δ values is presented by Hill (1950). Friction can be easily introduced into the analysis. From what has already been learned, however, its influence on the pressure must enter more at lower values of Δ, since friction has been found to have no effect on the pressure for nonpenetrating deformation at high Δ.

Deformation efficiency and redundant strains are other parameters which also depend strongly on the shape coefficient. As pressure tends to increase with Δ, the deformation efficiency tends to fall. The fall is due to the redundant strains which are defined quantitatively by the redundant-strain factor.

It is through these three parameters, namely, the tool pressure, the deformation efficiency and the redundant strain, that the shape coefficient affects material behaviour in all metal forming processes and it can be an important criterion for choosing the optimal mathematical model. The effect of Δ on forming conditions was extensively investigated by Wright (1976) for wire and rod drawing, by Blazynski (1979) and Skolyszewski (1986) for tube drawing and by Tselikov and Grishkov (1970), Pietrzyk (1983b,1986) and Pietrzyk and Kusiak (1986) for flat rolling.

The shape coefficient in the rolling process is defined as the average thickness to average length ratio in the deformation zone and is given as

$$\Delta = \frac{h_{av}}{l_d} = \sqrt{\frac{h_1}{R'r}\frac{(2-r)}{2}} \tag{5.1}$$

where h_{av} is the average thickness, h_1 is the initial thickness, l_d is the length of the deformation zone, R' is the radius of the flattened roll and r represents the reduction in a pass.

The values of the shape coefficient Δ calculated for a practical range of flat rolling parameters are given in Table 5.1. Correlation between the Δ coefficient and the characteristics of material flow in the flat rolling process is discussed below.

5.1. Correlation Between the Shape Coefficient and Load Parameters

5.1.1 The Effect of Δ on the Friction Stresses

The correlation between the distribution of friction stresses and the shape coefficient has been investigated by Tselikov and Grishkov (1970). They assumed that there are three possibilities of frictional conditions in the deformation zone, namely:

(a) slipping with friction stresses given by equation (1.7),
(b) sticking, when the friction stress equals the yield strength in shear k,
(c) so called "dead zone", when the difference between the roll and the strip velocity is small and stresses are within the elastic range.

The length of the dead zone may be given as

h_1/R	reduction							
ratio	0.05	0.10	0.15	0.20	0.25	0.30	0.35	0.40
1	4.36	3.00	2.39	2.01	1.75	1.55	1.39	1.26
0.5	3.08	2.12	1.69	1.42	1.24	1.1	0.97	0.89
0.2	1.95	1.34	1.07	0.90	0.78	0.69	0.62	0.57
0.1	1.38	0.95	0.76	0.64	0.55	0.49	0.44	0.40
0.05	0.98	0.67	0.53	0.45	0.39	0.35	0.31	0.28
0.02	0.62	0.42	0.34	0.28	0.25	0.22	0.20	0.18
0.01	0.44	0.30	0.24	0.20	0.18	0.16	0.14	0.13

Table 5.1 Values of the shape coefficient Δ calculated for various rolling parameters.

$$l_3 = nh_{av} \tag{5.2}$$

where n is a coefficient which, according to Tselikov and Grishkov, should be between 0.3 for low friction coefficients and 0.9 for large friction coefficients.

The pattern of friction stresses depends strongly on the shape coefficient Δ. For small values of Δ, approximately below 0.2, all three friction patterns appear in the deformation zone, as shown in Fig.5.1a. For larger values of Δ, between 0.2 and 0.5, the deformation zone is too short for friction stresses to reach the yield strength in shear, k, and the sticking zone does not exist, see Fig.5.1b. Further increase of the shape coefficient is connected with thicker strips and shorter deformation zones. Both these factors lead to an increase of the dead zone according to equation (5.2), which can be seen in Fig.5.1c where the distribution of friction stresses for Δ between 0.5 and 2 is shown. Finally, when Δ exceeds 2 the dead zone covers the whole arc of contact (Fig.5.1d) and the deformation does not penetrate across the whole section of the strip. Of course, the boundaries between these ranges of the shape coefficient are not sharp and they may vary slightly depending on other rolling parameters.

Tselikov and Grishkov's (1970) model of friction stresses is based on a somewhat artificial division of the deformation zone and it is not very convenient for numerical modelling. Similar results can be obtained by the introduction of one function which describes the distribution

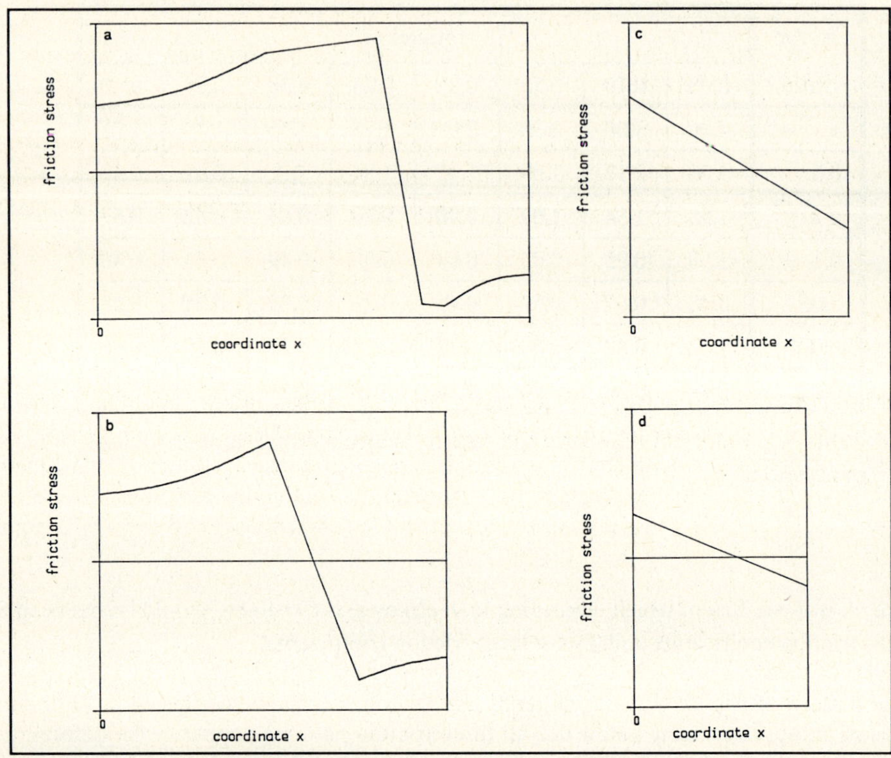

Fig.5.1 Friction stress distributions for various shape coefficients; a: $\Delta \leq 0.2$; b: $0.2 < \Delta \leq 0.5$; c: $0.5 < \Delta \leq 2$; d: $\Delta > 2$.

of friction stresses in the deformation zone, as shown by Polukhin (1975)

$$\tau = \mu p \operatorname{sign}(l_b - x) \left| \frac{l_b - x}{l_b} \right|^{\nu} \qquad \text{for backward slip zone} \tag{5.3}$$

and

$$\tau = \mu p \operatorname{sign}(l_b - x) \left| \frac{l_d - x}{l_d - l_b} \right|^{\nu} \qquad \text{for forward slip zone} \tag{5.4}$$

where l_a is the length of the arc of contact, l_b is the length of the backward slip zone, μ is the constant value of the coefficient of friction, p is the roll pressure and x represents the coordinate along the strip axis.

In equations (5.3) and (5.4) ν is a constant between 0 and 1, depending on the shape coefficient Δ. The relationship between ν and Δ was investigated experimentally by Derkach et al. (1968) for larger values of the shape coefficient relevant for hot flat rolling and the following function was proposed

$$\nu = 1.5\Delta + 0.175\frac{1}{\Delta} - 1 \tag{5.5}$$

Distributions of the relative friction stress $\tau / \mu p$ for various values of coefficient ν are shown in Fig.5.2. It can be seen from this figure that an introduction of relationships (5.3), (5.4) and (5.5) gives results which are in qualitative and quantitative agreement with Tselikov's model, however, the friction stresses are described by continuous functions. Moreover, relationships (5.3), (5.4) and (5.5) are very convenient in numerical solutions of von Karman's equation.

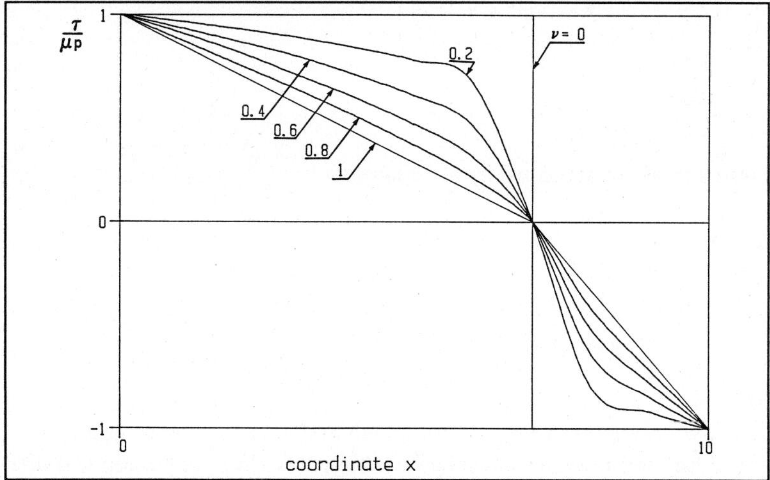

Fig.5.2 Distribution of the relative friction stress along the arc of contact.

5.1.2 The Effect of Δ on Forces and Torques

Values of the shape coefficient in flat rolling processes are usually between 0.1 and 2. In this range the effect of Δ on the roll pressure is through the friction stresses only - see Backofen (1972) and Pietrzyk (1986). In general, for greater values of Δ the friction hill decreases and the magnitude of the roll pressure approaches that of the yield strength. As shown by Tselikov and Grishkov (1970), Ford and Alexander (1963) and Pietrzyk (1983) the roll pressure may be expressed as a function of only one geometrical parameter: Δ. A typical example of that approach is the simplified formula for the roll force in hot rolling, suggested by Ford and Alexander (1963)

$$F = \bar{k}\, l_d \left(1.571 + \frac{1}{2\Delta} \right)$$

where $\bar{k}$ is the average yield strength in shear and l_d is the length of the arc of contact.

The shape coefficient's role in the calculations of the rolling torque is also of some importance. There are two methods by which the torque may be evaluated. One is based on the integration of the friction stresses along the arc of contact according to the following formula

$$M = R \int_0^{l_d} \tau\, dx \tag{5.6}$$

and the second method is described by the equation

$$M = \psi\, l_d F \tag{5.7}$$

where F is the roll force per unit width, R is the roll radius, l_d is the length of the arc of contact and ψ is the coefficient of the lever arm.

The coefficient ψ is usually difficult to determine and its value is often assumed depending on the type of the rolling mill used. It was shown by Pietrzyk (1983) that the coefficient of the lever arm depends closely on the shape coefficient Δ. This conclusion was arrived at on the basis of both experimental investigations and analyses of theoretical formulae.

The roll torque can be also calculated by the upper bound approach (Avitzur, 1968). The basic equation which describes the power balance in the deformation zone is (Pietrzyk, 1983)

$$M\omega = \int_V \sigma_i \dot{\epsilon}_i dV + \int_{S_1} \frac{\sigma_p}{\sqrt{3}} |v_{yo}| dS_1 + \int_S \tau |v_s| dS \tag{5.8}$$

where V is the volume of the deformation zone, S_1 represents the surface of the velocity discontinuity, S is the contact surface, M is the roll torque, ω is the roll angular velocity, σ_i and $\dot{\epsilon}_i$ are the effective stress and the effective strain rate respectively, σ_p is the yield strength, v_{yo} represents a velocity discontinuity, τ is the shear stress at the contact surface and v_s is the slip velocity. Introducing simplifications suggested by Tselikov and Grishkov (1970), i.e. substituting the arc of contact by its chord and assuming uniform flow of metal, allows the derivation of the power components in equation (5.8), as follows:

- power for plastic deformation

$$\dot{W}_p = \int_V \sigma_i \dot{\epsilon}_i dV = \frac{2}{\sqrt{3}} \sigma_p v_1 b h_1 \ln(1-r) \tag{5.9}$$

- power for friction losses

$$\dot{W}_f = \int_S \tau |v_s| dS = b v_1 \tau [\frac{l_d}{r} \ln \frac{(l_d - r l_n)^2}{(1-r) l_d^2} + \omega R (2 l_n - l_d)] \tag{5.10}$$

- power dissipated due to the velocity discontinuities

$$\dot{W}_s = \int_{S_1} \frac{\sigma_p}{\sqrt{3}} |v_{yo}| dS_1 = \frac{2}{\sqrt{3}} b \sigma_p v_1 r l_d \left(\frac{\Delta}{2-r}\right)^2 \tag{5.11}$$

In equations (5.9), (5.10) and (5.11) b is the width of the strip, v_1 is the strip entry velocity, R is the roll radius, l_d is the length of the arc of contact, l_n is the length of the forward slip zone and r represents the reduction per pass. By substituting equations (5.9), (5.10) and (5.11) into relationship (5.8) and using Tselikov's formula (3.47) for the roll separating force the lever arm coefficient is calculated as

$$\psi = \frac{\frac{1}{\sqrt{3}}A\ln(1-r) + \left\{\mu\frac{1}{r}\ln\left[(1-r)\left(\frac{h_n}{h_2}\right)^2\right] + \frac{1}{r}\left[\frac{h_1}{h_n}(2-r)-2\right]\right\} + A^2\frac{r}{\sqrt{3}}}{2.3A^2\left[\left(\frac{h_1}{h_n}\right)^\delta + \left(\frac{h_n}{h_2}\right)^\delta - 2\right]\frac{r}{\mu}} \tag{5.12}$$

where

$$A = \frac{\Delta}{2-r}$$

$$\delta = 2\mu\frac{l_d}{h_1 r} = \mu\frac{2-r}{\Delta r}$$

$$\frac{h_n}{h_2} = \left[\frac{1+\sqrt{1+(\delta^2-1)\left(\frac{h_1}{h_2}\right)^\delta}}{\delta+1}\right]^{\frac{1}{\delta}}$$

$$\frac{h_n}{h_1} = \frac{h_n}{h_2}(1-r)$$

In the above equations h_1 and h_2 represent the initial and final thickness of the strip, respectively, and h_n is the thickness at the neutral point.

Calculations carried out by Pietrzyk (1983b) using (5.12) show that the influence of the reduction r and friction coefficient μ on the lever arm coefficient is negligible. Variations of reduction between 5% and 60% and variations of friction coefficient between 0.25 and 0.35 result in lever arm coefficient variations of less than 2%. It can be concluded then, that the shape coefficient Δ is the single significant rolling parameter which affects the value of the coefficient ψ. Close dependence between the lever arm coefficient and the shape coefficient is also obtained from the experimental results presented in Fig.5.3 which shows the values of ψ calculated from roll forces and torques measured on various hot rolling mills. A reasonably good correlation between ψ and Δ is observed in that figure and the relationship between the two coefficients can be approximated by

$$\psi = 0.594\Delta^{0.3146} \tag{5.13}$$

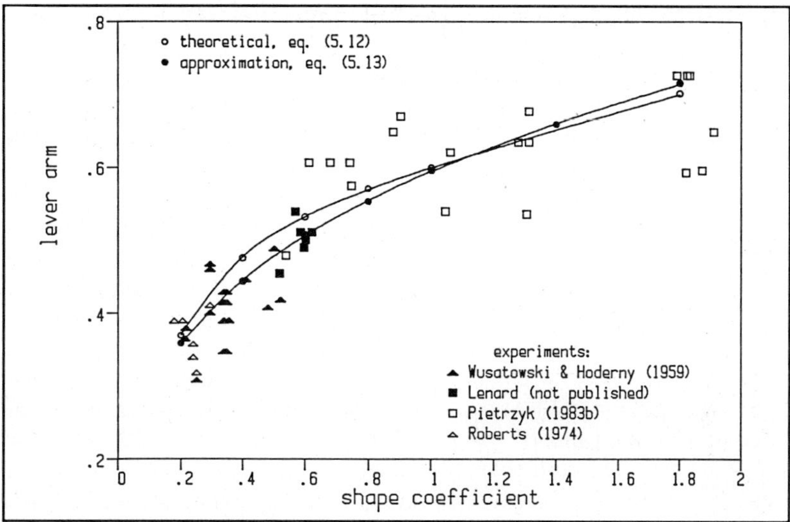

Fig.5.3 Lever arm vs. shape coefficient relationship.

A similar correlation between the lever arm coefficient and the shape coefficient Δ in the hot flat rolling process was observed by Andrade et al. (1987), who suggested the following relationship

$$\ln \psi = 0.166 \ln \frac{1}{\Delta} - 0.683 \tag{5.14}$$

Simple rearrangement of equation (5.14) leads also to a power function of the type given by relationship (5.13) but with slightly different coefficients.

All of the above investigations emphasize the close correlation between the shape coefficient and the lever arm coefficient. Since it is a closed form formula which does not require numerical integration, relationship (5.13) can be very useful in calculations of the roll torque in the hot flat rolling process. Similar analysis can also be done for the cold rolling process which involves lower values of shape coefficients, however, due to the significant influence of the flattening of the roll on the lever arm coefficient a consistent presentation of the results becomes more difficult.

5.2. The Effect of the Shape Coefficient on Strain and Strain Rate Distributions

The distributions of the strains and strain rates are also closely correlated with the shape coefficient. Detailed analysis of the influence of the shape coefficient on the strain rate and strain fields in the roll gap was carried out by Pietrzyk (1986) and Pietrzyk and Kusiak (1986). Some typical results of that analysis are presented below.

5.2.1 Finite Element Analysis of Strain Rate Fields for Various Shape Coefficients

A rigid-plastic finite element approach coupled with a finite element analysis of the thermal events described in Sections 4.1 and 4.2 was used to simulate strain rate and strain distributions during rolling with various shape coefficients and the results of the calculations are presented in Figs 5.4, 5.5 and 5.6. Since the coefficient Δ depends on the reduction r and on the h_1/R' ratio assigned by equation (5.1), to make the comparison of results possible the calculations were carried out for one value of r = 0.2 and for four values of h_1/R', specifically, 0.01, 0.063, 0.249 and 0.556. This resulted in the values of shape coefficients of 0.2, 0.5, 1.0 and 1.5, respectively. Typical results of calculations of the effective strain rate distribution for various values of Δ are shown in Fig.5.4. Analysis of the isoclines in Fig.5.4 shows that the maximum values of the strain rate decrease with Δ. Also, for large shape coefficients, a wide area of very low strain rates appears close to the neutral point. This phenomenon supports Tselikov and Grishkov's (1970) theory which suggests that a dead zone exists near the neutral point, the length of which increases with Δ. Tselikov's simulation of the friction stresses was somewhat artificial since he assumed that $\tau = k$ in the sticking zone and, further, a linear change of τ close to the neutral point. In the present finite element solution the physical simulation of the interfacial shear stress is achieved by an introduction of the velocity dependent friction forces given by equation (4.14). As a result a smooth curve of friction stress distribution is obtained and, moreover, a sticking zone around the neutral point is simulated realistically. As seen in Fig.5.4, the length of the sticking zone increases with Δ, in qualitative agreement with the classical theory of rolling described, among the others, by Orowan (1943), Wusatowski (1969) and Tselikov and Grishkov (1970).

Fig.5.5 shows the distribution of the shear strain rate in the deformation zone. It can be seen from this figure that negative strain rates in the entry zone increase significantly with Δ. This situation is reversed further inside the deformation zone, where maximum positive values of $\dot{\epsilon}_{xy}$ appear for Δ = 0.2. The zone of the negative shear strain rates close to the exit from the roll gap decreases with increasing Δ and, finally, it disappears for Δ larger than unity. The $\dot{\epsilon}_{xy}/\dot{\epsilon}_x$ ratio is always lower for smaller shape coefficients which is observed in the

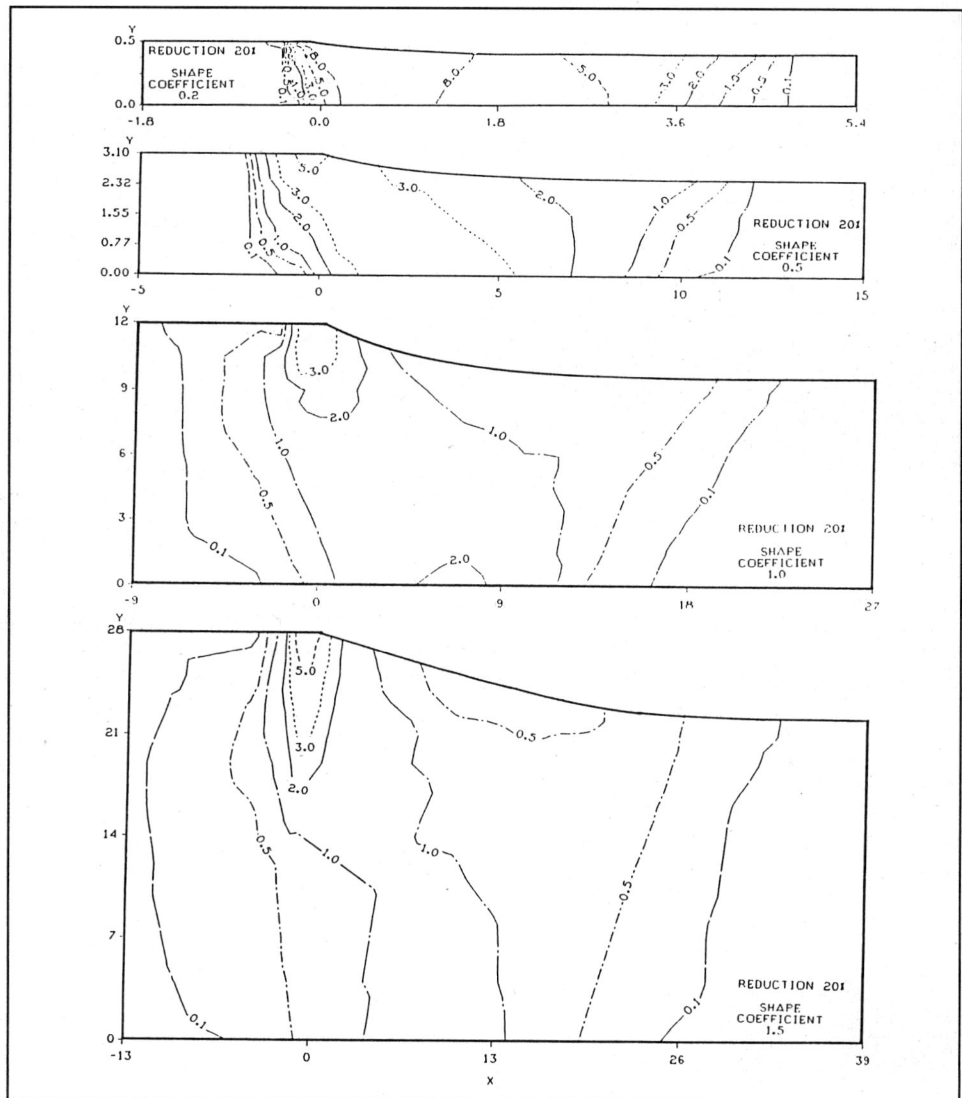

Fig.5.4 Distribution of the effective strain rate in the roll gap for various shape coefficients.

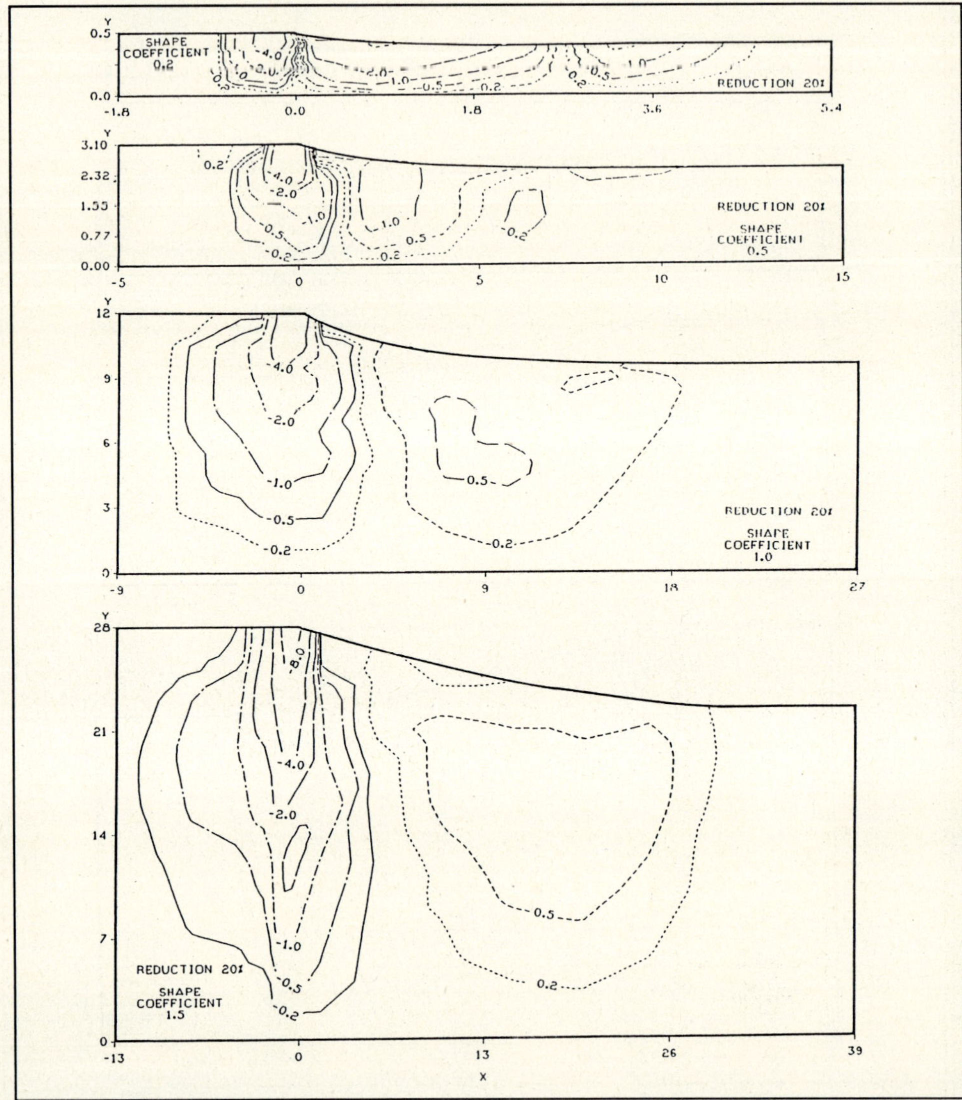

Fig.5.5 Distribution of the shear strain rate in the roll gap for various shape coefficients.

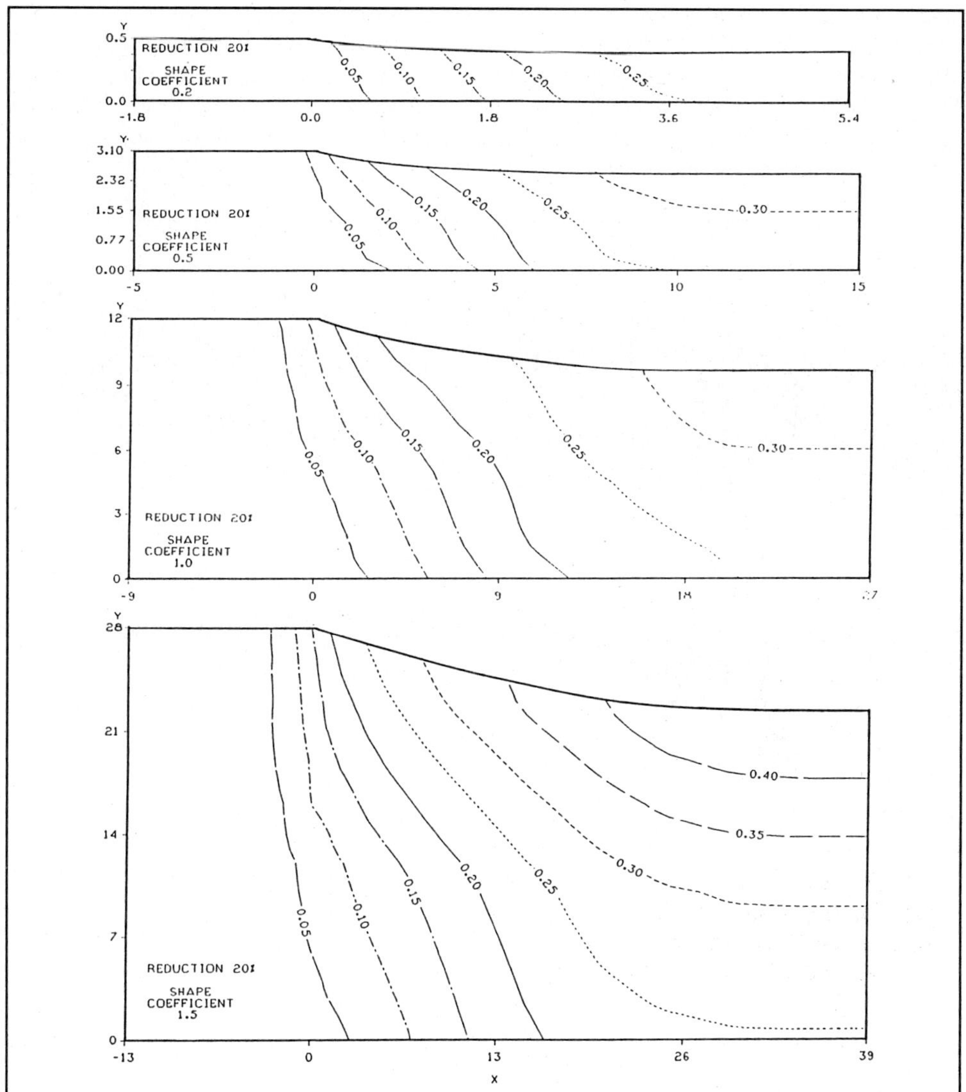

Fig.5.6 Distribution of the effective strain in the roll gap for various shape coefficients.

analysis of the effective strain distribution in the deformation zone, shown in Fig.5.6. Non-uniform strains observed in this figure lead to the effective strain distribution along the material thickness after the pass which is presented in Fig.5.7. It is observed that the effective strains increase with Δ. It should be pointed out that strain variations along the material thickness lead to gradients of structural and mechanical properties in all materials that strainharden or undergo any strain dependent structural change.

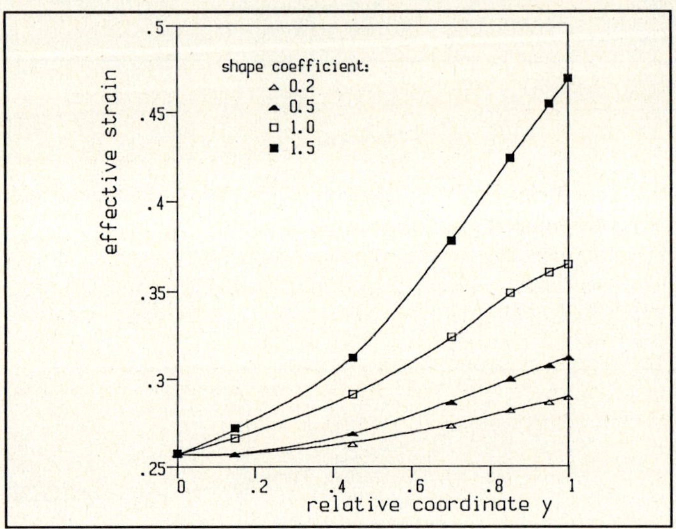

Fig.5.7 Effective strain distribution across the strip after rolling.

5.2.2 Efficiency of the Rolling Process

Efficiency of deformation in various metal forming processes is often measured by the relative amounts of shear strains, which are usually called redundant strains. Quantitative assessment of the redundant strains can be done in different ways, one of which is based on a coefficient called redundant strain factor and defined as

$$\Phi = \frac{\bar{\epsilon}_i}{\epsilon_{iu}} \tag{5.15}$$

where $\bar{\epsilon}_i$ is the average effective strain of the cross section and ϵ_{iu} is the effective strain for uniform deformation.

Finite element calculations of strain distribution in the roll gap allow the determination of the redundant strain factor with reasonable accuracy. The relationship between this factor and the shape coefficient is shown in Fig.5.8. The increase of Φ with Δ is due to the increase of the shear strain rates related to the effective strain rate. The influence of the shear strain rates is measured by the coefficient κ defined as

$$\kappa = \frac{\dot{\epsilon}_{xy}}{\dot{\epsilon}_i} \tag{5.16}$$

The value of this coefficient varies in the volume of the deformation zone, being equal to zero along the axis of symmetry and achieving maximum in the area close to the initial point of contact between the strip and the roll. Maximum values of κ can be used as a measurement of the efficiency of the rolling process, the lower κ being connected with more efficient deformation. The effect of the shape coefficient on the maximum κ values, which appear

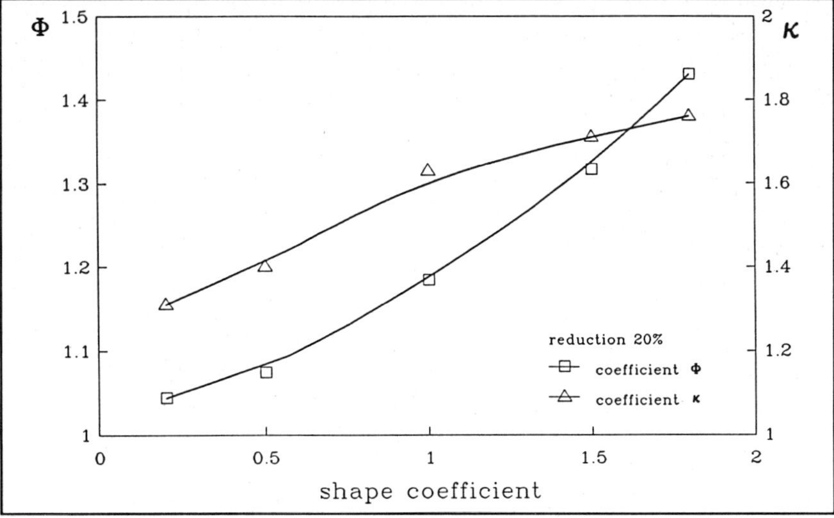

Fig.5.8 Relationships between factors Φ and κ and the shape coefficient

in the element close to the point of bite, is also presented in Fig.5.8. Significant increase of the coefficient κ with Δ is observed in this figure, indicating decrease of the efficiency of deformation. It should be pointed out, however, that this analysis does not include the friction, influence of which increases at lower values of Δ.

Strain rate and strain variations along the deformation zone are shown in Figs 5.9, 5.10, 5.11 and 5.12. Separate graphs are presented for the areas close to the roll-workpiece interface (Figs 5.10 and 5.12) and for the strip axis (Figs 5.9 and 5.11). The figures confirm the existence of the dead zone around the neutral point, the length of the zone increasing for increasing values of Δ. This is demonstrated in Fig.5.10 where, for larger shape coefficients, the effective strain rates close to the contact surface are almost equal to zero in the major part of the roll gap, indicating no slip conditions between the roll and the strip. The graphs plotted in Figs 5.9 and 5.10 show that, for the same rolling velocities, the effective strain rates are generally much higher for lower shape coefficients. This situation is due to a shorter time of deformation for processes with lower Δ. It is obvious that larger strain rates are necessary to achieve the same strain in shorter time. In consequence, the absolute values of the strain rates are not convenient in comparative analyses of the efficiency of deformation. Relative redundant strains represented by the coefficient κ, given by equation (5.16), are more useful for comparisons. Analysis of Fig.5.11 shows that the deformation pattern in the centre of the sample does not depend on the shape coefficient and the final effective strain in that area is close to the strain in a uniform deformation. The influence of the redundant strains is noticeable in Fig.5.12 which shows the situation close to the strip surface. Final effective strains in that area appear to be much larger for larger shape coefficients.

Analysis of the results presented in Figs 5.3 - 5.8 shows that strain inhomogeneity increases with Δ. This in turn will cause the stress distribution to be nonhomogeneous in the deformation zone as well. It is recalled that almost all rolling theories are based on the slab analysis which assumes a uniform flow of metal. The above considerations suggest that they should not be applied to rolling processes with high values of Δ. Limiting value of the shape coefficient, below which strain and stress inhomogenities can be neglected, is assessed to be about 0.6 - 0.7. Analyses also show that a wide dead zone appears close to the neutral points in rolling processes with large Δ values. Length of the dead zone increases with Δ, the length of the total plastic zone measured along the material axis having an opposite tendency.

The discussion presented in this section and earlier investigations of Pietrzyk (1986) and Pietrzyk and Kusiak (1986) deal with the efficiency of pure plastic deformation during rolling and the friction losses are not included in the analyses. The effect of friction on the efficiency of the metal forming processes for various shape coefficients has been investigated by several

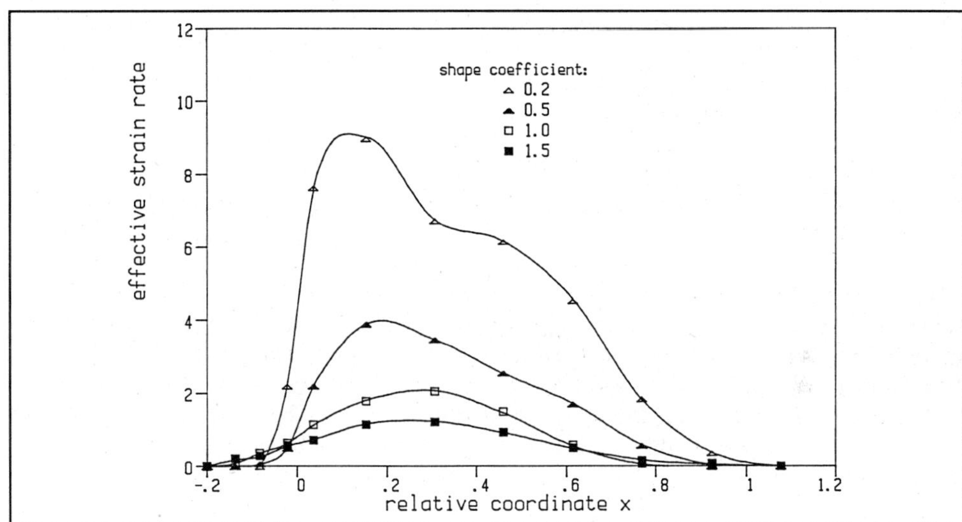

Fig.5.9 Distribution of the effective strain rate in the centre of the sample along the roll gap.

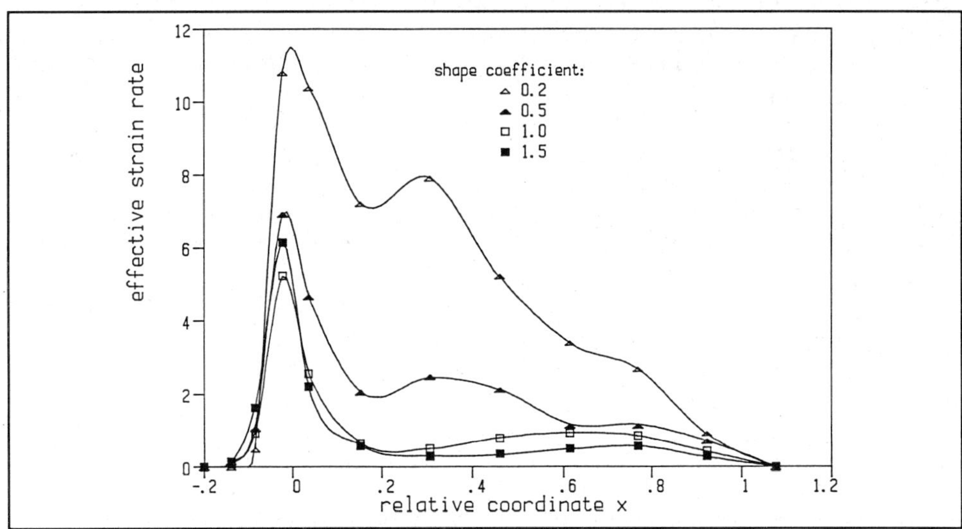

Fig.5.10 Distribution of the effective strain rate close to the surface along the roll gap.

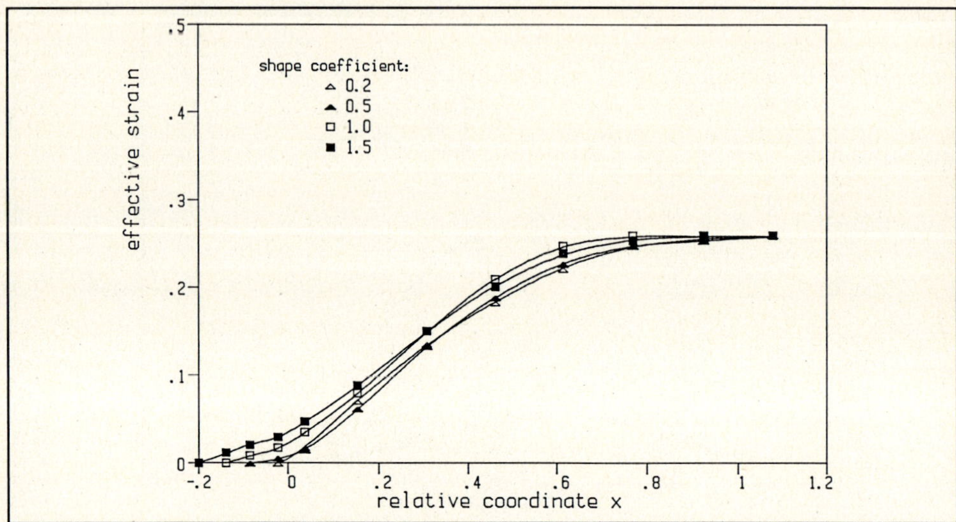

Fig.5.11 Distribution of the effective strain in the centre of the sample along the roll gap.

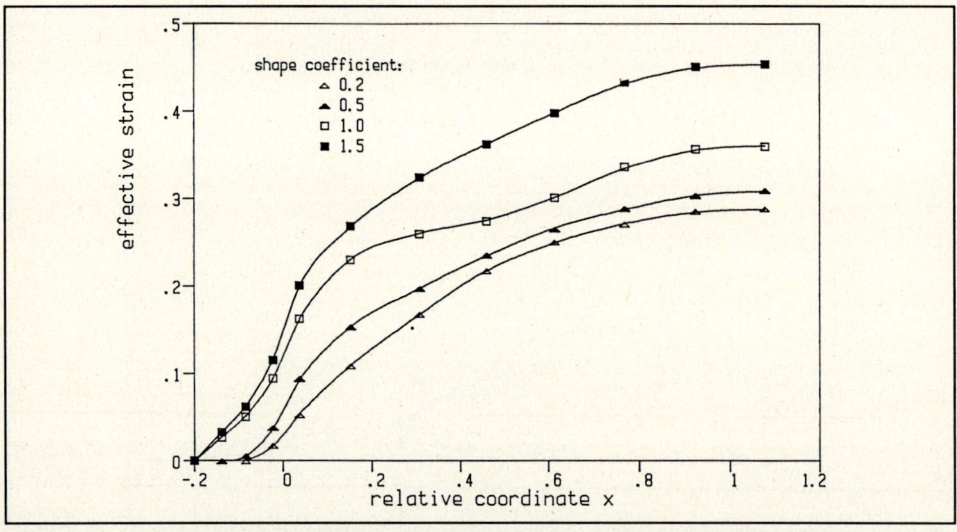

Fig.5.12 Distribution of the effective strain close to the surface along the roll gap.

authors - see e.g. Backofen (1972), Tselikov and Grishkov (1970) and Jaglarz et al. (1980) - and it was shown that the influence of friction is larger for smaller shape coefficients. In consequence, in strip rolling processes connected with low shape coefficients the friction coefficient should be decreased. In blooming mills where the shape coefficients are large and the influence of friction on the loads is negligible, friction should be increased in order to aid entry.

CHAPTER 6

CONCLUSIONS

Mathematical modelling of the flat rolling process has been considered in this book with the emphasis placed on experimental substantiation of the predictions of the models. A general formulation of the problem has been presented in Chapter 1. In Chapter 2, phenomena, common to both one and two-dimensional analyses, have been discussed, these being the interfacial forces during rolling and the material's resistance to deformation. The traditional models, based on equilibrium considerations, have been treated next. Thermal-mechanical modelling and a description of the role of the shape coefficient are included in the following chapters.

The publications of concern here may be divided into three major groups. The first one deals with the formulation and the solution of the governing equations of the processes. The next considers the development of experimental techniques and the generation of data. Finally there are the studies that develop and compare the measurements and the predictions. The information presented in projects in each of these groupings is of primary importance.

A general conclusion has emerged from the research which led to the publication of the present book, concerning the success of a predictive model of metalforming and specifically, of the flat rolling process. This has to do with the proliferation of mathematical models which concentrate on methods of treatment of the equations, efficient computer schemes, ways to decrease memory requirements and elegant procedures over those investigating the events occurring on the boundaries of the domains of interest. It is fairly easy and straightforward to formulate the field equations dealing with mechanical and thermal equilibrium. Solutions are of course not so simple. Most of the difficulties are encountered however, when the initial and boundary conditions are considered. For mechanical systems these involve the properties of the material being rolled, the forces operating at the roll/metal interface and the shape of that interface. In thermal modelling it is the heat transfer coefficient that needs to be described.

The other point of importance is the problem of substantiation of the predictions of the models. There is a noticeable scarcity of publications dealing with experiments in metal-forming. It is understood that these tests are difficult and time consuming to perform. They need specialized and expensive equipment, data acquisition systems, materials and an understanding of several disciplines, including mechanics, heat transfer, metallurgy, tribology, and experimental techniques. The authors would like to underline the pressing need for the data however, since only through a thorough investigation of the differences between measurements and predictions could either of them be improved.

It may also be concluded that accurate mathematical description of interfacial frictional conditions and heat transfer would enhance the predictive capabilities of the models. Use of data from a set of preliminary experiments backs up the above statement. The need for further extensive testing is strongly emphasized, however. The dependence of both frictional conditions and the heat transfer coefficient on process parameters and material properties could then be more fully understood and their choice in modelling would be eased.

As far as the material's resistance to deformation is concerned, further caution is recommended. Testing techniques and the manner of reporting flow strength data should be standardized. The pre-test thermal treatment and austenite grain size, the sample size and end geometry, the configuration and the material of the loading ram, the precise time schedule of the experiment, the lubricant, the temperature measuring system, the type of test and the material's chemical composition should always be reported.

Examination of the predictive capabilities of one dimensional models led to the observation that all of those considered are of adequate accuracy. Statistical comparison to production data - the real test of any model - indicated that accuracy and consistency are directly related to mathematical rigour.

Thermal-mechanical modelling reinforced earlier conclusions regarding the effect of the heat transfer coefficient on the predictions of the temperatures, forces and torques. The value of α used in the computations should be chosen carefully, including considerations of how it was determined and calculated from the data. In laboratory conditions, assuming constant roll temperatures 4800 W/m^2K was appropriate; if the model accounted for the increasing roll temperature, 13000 W/m^2K gave the best results. Modelling events on a hot strip mill required α = 50000 W/m^2K for accuracy. This number however, is expected to be very strongly dependent on the particular mill configuration and on rolling practices.

Two versions of the model were used in the computations. One described steady state conditions while the other was concerned with transient behaviour. In most cases the steady state version was found to be appropriate.

Considering the broad ranges of the heat transfer coefficient, changing its value was expected to cause significant differences in the predicted distributions of the temperature. While changes were undeniably present, they were not as large as predicted. Increasing the roll/strip heat transfer coefficient by more than an order of magnitude resulted in an approximately 150°C change in the computed temperatures.

A general conclusion - one that has often been stated by steel mill engineers in the past - has also been reinforced. This indicates that it is difficult to take results from a single stand, small scale mill and expect them to be valid for full scale industrial equipment. The possibility of developing appropriate scaling factors should be investigated.

Concerning the question: Which model should be used under a specific set of circumstances? - a preliminary answer may be given, subject, of course to corrections as computational and experimental techniques improve and more data are generated. If hot strip rolling is considered and only a general "feel" for roll forces and torques is needed, Sims' method, using average temperatures in the pass should be satisfactory. If thermal-mechanical processing of microalloyed steels, including adaptive control of the strip mill and predictions of the evolution of the microstructure are required, use of at least a two dimensional thermal-mechanical model is recommended. The length of computations in such a model of course preclude its use in on-line situations. Research to overcome this problem - perhaps by precalculating, storing and retrieving numbers as required - should be undertaken.

REFERENCES

Alexander, J.M., 1972, On the Theory of Rolling, Proc. R. Soc. Lond., **A.326**, pp. 535-563.

Al-Rubeye, H., 1980, Wear and Temperature in Sliding Contact, ASME, J. Lubr. Techn., **102**, pp. 107-112.

Al-Salehi, F.A.R., Firbank, T.C. and Lancaster, P.B., 1973, An Experimental Determination of the Roll Pressure Distribution in Cold Rolling, Int. J. Mech. Sci., **15**, pp. 693-710.

Altan, T. and Boulger, F.W., 1973, Flow Stress of Metals and its Application in Metal Forming Analysis, ASME, J. Eng. Ind., **95**, pp. 1009-1019.

Andrade, H.L., Roldan, A.J., Pereira, J.E.A. and Hauck, G.A.C., 1987, A Mathematical Model for Rolling Schedule Design and Computer Control of the USIMINAS 4100 mm Plate Mill, Proc. 4th Int. Steel Rolling Conf., Deauville, vol.1, pp. C7.1-C7.7.

Armitage, B., McCutcheon, B.B. and Newton, L.D., 1976, Power Requirements for Rolling Columbium Bearing, HSLA Steel, Proc. 18th Conf. Mech. W. and Steel Proc., Harvey, Ill., pp.13-34.

Atreya, A. and Lenard, J.G., 1979, A Study of Cold Strip Rolling, ASME, J. Eng. Mat. Techn., **101**, pp. 129-134.

Avitzur, B., 1968, Metal Forming - Processes and Analysis, McGraw Hill.

Avitzur, B. and Nowakowski, A., 1980, Analysis of Strip Rolling as an Adiabatic Process, Proc. 21st Int. MTDR Conf., Swansea, vol.1, pp. 29-32.

Backofen, W.A., 1972, Deformation Processing, Addison-Wesley.

Bacroix, B., Akben, M.G. and Jonas J.J., 1981, Effect of Molybdenum on Dynamic Precipitation and Recrystallization in Niobium and Vanadium Bearing Steels, Proc. Conf. Thermomechanical Processing of Microalloyed Austenite, DeArdo, A.J., Ratz, G.A. and Wray, P.J., (eds), Metallurgical Society of AIME, Pittsburgh, pp. 293-318.

Banerji, A. and Rice, W.B., 1972, Experimental Determination of Normal Pressure and Friction Stress in the Rolling Gap During Cold Rolling, Annals of CIRP, **21**, pp. 53-54.

Bernick, L. and Wandrei, C.L., 1978, Friction in Hot Rolling of Steel, Proc. 1st Int. Conf. on Lubrication Challenges in Metalworking Processes, Chicago, pp. 71-80.

Bertrand-Corsini, C., David, C., Bern, A., Montmitonnet, P., Chenot, J.L., Buessler, P. and Fau, F., 1988, A Three Dimensional Thermomechanical Analysis of Steady Flows in Hot Forming Processes. Application to Hot Flat Rolling and Hot Shape Rolling, Modelling of Metal Forming Processes, Chenot, J.L. and Onate, E. (eds), Kluwer Academic Publishers, pp. 271-279.

Bland, D.R. and Ford, H., 1948, The Calculation of Roll Force and Torque in Cold Strip Rolling with Tensions, Proc. Inst. Mech. Eng, **159**, pp. 144-153

Blazynski, T.Z., 1979, Geometry Factors and Inhomogeneity of Flow in Rotary Piercing and Cold Drawing of Seamless Tubing, Int. J. Mech. Sci., **21**, pp. 527-536.

Boër, C.R., Rebelo, N., Rydstad, H. and Schröder, G., 1986, Process Modelling of Metal Forming and Thermomechanical Treatment, Springer-Verlag, Berlin.

Britten, D.D. and Jeswiet, J., 1986, A Sensor Measuring Normal Forces with Through and Transverse Friction Forces in Roll Gap, Trans. NAMRC XIV, pp. 355-358.

Bryant, G.F. and Chiu, T.S.L., 1982a, Simplified Roll-Temperature Model: Convective Cooling, Metals Technol., **9**, pp. 478-484.

Bryant, G.F. and Chiu, T.S.L., 1982b, Simplified Roll-Temperature Model: Spray-Cooling and Stress Effects, Metals Technol., **9**, pp. 485-492.

Bryant, G.F. and Heselton, M.O., 1982, Roll-Gap Temperature Models for Hot Mills, Metals Technol., **9**, pp. 469-477.

Buessler, P. and Schoenberger, P., 1987, Analysis of Hot Rolling by an Elasto-Visco-Plastic Finite Element Method, Proc. 4th Int. Steel Rolling Conf., Deauville, vol. 2, pp. F8.1-F8.7.

Carslaw, H.S. and Jaeger, J.C., 1959, Conduction of Heat in Solids, Oxford University Press, London.

Cerni, S., Weinstein, A.S. and Zorowski, C., 1963, Temperatures and Thermal Stresses in the Rolling of Metal Strip, Iron and Steel Eng. Yearbook, pp. 165-171.

Chen, C.C. and Kobayashi, S., 1978a, Deformation Analysis of Multi-Pass Drawing and Extrusion, Annals of the CIRP, **27**, pp. 151-155.

Chen, C.C. and Kobayashi, S., 1978b, Rigid-Plastic Finite-Element Analysis of Ring Compression, Application of Numerical Methods to Forming Processes, ASME, ADM - vol.28, pp .163-174.

Cole, I.M. and Sansome, D.H., 1968, A Review of the Application of Pin Load Cell Pressure Measurement Techniques to Metal Deformation Processes, MTDR, pp. 271-286.

Cook, M. and Larke, E.C., 1945, Resistance of Copper and Copper Alloys to Homogeneous Deformation in Compression, J. Inst. Metals, **71**, pp. 371-390.

Dadras, P. and Wells, W.R., 1984, Heat Transfer Aspects of Nonisothermal Axisymmetric Upset Forging, ASME, J. Eng. Ind., **106**, pp. 187-195.

Dawson, P.R., Thompson, E.G., 1978, Finite Element Analysis of Steady State Elasto-Visco-Plastic Flow by the Initial Stress Rate Method, Int. J. Num. Meth. Eng., **12**, pp. 47-57.

DeArdo, A.J., 1988, Accelerated Cooling: A Physical Metallurgy Perspective, Canadian Metalurg. Quart., **27**, pp. 141-154.

Derkach, W.A., Zaykov, M.A. and Tseluykov, W.S., 1968, Chislenniy Analiz Napriazhennowo Sostayaniya Prokatyvayemowo Metalla s Pomoshchyu ECWM, Izv. VUZ, Chern. Metal., no.10, pp. 96-101.

Devadas, C. and Samarasekera, I.V., 1986, Heat Transfer During Hot Rolling of Steel Strip, Ironmaking and Steelmaking, **13**, pp. 311-321.

D'Orazio, L.R., Mitchell, A.B. and Lenard, J.G., 1985, On the Response of Nb Bearing HSLA Steels to Single and Multistage Compression, Proc. Conf. HSLA Steels'85, Beijing, publ. by ASM, Metals Park, Ohio, pp. 1-9.

Douglas, J.R. and Altan, T., 1975, Flow Stress Determination for Metals at Forging Rates and Temperatures, ASME, J.Eng. Ind., **97**, pp. 66-76.

El-Kalay, A.K.E.H.A. and Sparling, L.G.M., 1968, Factors Affecting Friction and their Effect Upon Load, Torque and Spread in Hot, Flat Rolling, J. Iron Steel Inst., **197**, pp. 152-163.

Ellwood, S., Griffiths, L.J. and Parry, D.J., 1982, Materials Testing at High Constant Strain Rates, J. Phys. E.: Sci. Instrum., **15**, pp. 280-282.

Engineering Properties of Steel, 1982, publ. by ASM, Metals Park, Ohio.

Ford, H. and Alexander, J.M., 1963, Simplified Hot Rolling Calculations, J. Inst. of Metals, **92**, pp. 397-403.

Ford, H., Ellis, F. and Bland, D.R., 1951, Cold Rolling with Strip Tension, Part 1, A New Approximate Method of Calculation and a Comparison with other Methods, J. Iron Steel Inst., **168**, pp. 57-72.

Gallagher, R.H., 1975, Finite Element Analysis - Fundamentals, Prentice-Hall Inc., Englewood Cliffs.

Geleji, A., 1960-61, Bildsame ormung der Metalle in Rechnung und Versuch, Akademie Verlag, Berlin, pp. 227-233.

Gelin, J.C., Oudin, J. and Ravalard, Y., 1981, Determination of the Flow Stress-Strain Curves for Metals from Axisymmetric Upsetting, J. Mech. Work. Techn., **5**, pp. 297-308.

Ghosh, A.K., 1977, A Method for Determining the Coefficient of Friction in Punch Stretching of Sheet Metals, Int. J. Mech. Sci., **19**, pp. 457-470.

Gittins, A., Moller, R.H. and Everett, J.R., 1974, The Strength of Steels Under Hot Working Conditions, BHP Technical Bulletin, **18**, pp. 1-7.

Gittins, A., Everett, J.R. and Tegart, W.J.McG., 1977, Strength of Steels in Hot Strip Mill Rolling, Metals Technol., **4**, pp. 377-383.

Golten, J.W., 1969, Analysis of Cold Rolling with Particular Reference to Roll Deformation, Ph.D. dissertation, University of Wales, Swansea.

Hajduk, M., Zidek, M., Elfmark, J. and Kopec, S., 1972, Urceni strednich hodnot prirozenych deformacnich odporu nakladnich druhu oceli pro valcovani za tepla, Hutnicke listy, **27**, pp. 567-571.

Halal, A.S. and Kaftanoglu, B., 1984, Computer Aided Modelling of Hot and Cold Rolling of Flat Strip, Proc. Int. Conf. on Computers in Engineering, Las Vegas, pp. 120-126.

Harding, R.A., 1976, Ph.D. dissertation, University of Sheffield, Sheffield, UK.

Heinrich, J.C., Huyacorn, A.R., Mitchell, A.R. and Zienkiewicz, O.C., 1977, An 'Upwind' Finite Element Scheme for Two- Dimensional Convective Transport Equation, Int. J. Num. Meth. Eng., **11**, pp. 131-143.

Heinrich, J.C. and Zienkiewicz, O.C., 1977, Quadratic Finite Element Schemes for Two-Dimensional Convective Transport Problems, Int. J. Num. Meth. Eng. **11**, pp. 1831-1844.

Hill, R., 1950, The Mathematical Theory of Plasticity. Oxford University Press, London.

Hitchcock, J.H., 1935, Roll Neck Bearings, ASME Res. Publ.

Hsu, Y.W., 1980, Cold Rolling Force-Torque Slip Model, Proc. NAMRC VIII, Rolla, pp. 176-181.

Hwu, Y.-J. and Lenard, J.G., 1988, A Finite Element Study of Flat Rolling, ASME, J. Eng. Mat. Techn., **110**, pp. 22-27.

Jaglarz, Z., Leskiewicz, W. and Morawiecki, M., 1980, Technologia i urzadzenia walcowni wyrobow plaskich, Slask, Katowice.

Jarl, M., 1988, Friction and Forward Slip in Hot Rolling, Scand. J. Metal., **17**, pp. 2-7.

Jeswiet, J. and Rice, W.B., 1982, The Design of a Sensor for Measuring Normal Pressure and Friction Stress in the Roll Gap During Cold Rolling, Proc. NAMRC X, pp. 130-134.

Jeswiet, J. and Rice, W.B., 1975, Measurement of Strip Temperature in the Roll Gap During Cold Rolling, Annals of CIRP, **24**, pp. 153-155.

Jonas, J.J. and Sakai, T., 1984, A New Approach to Dynamic Recrystallization, Deformation Processing and Structure, Krauss, G., (ed.), ASM, Metals Park, pp. 185-228.

Jortner, D., Osterle, J.F. and Zorowski C.F., 1958, An Analysis of the Mechanics of Cold Strip Rolling, Iron Steel Eng. Yearbook, pp. 403-414.

Kaftanoglu, B., 1973, Determination of Coefficient of Friction under Conditions of Deep Drawing and Stretch Forming, Wear, **25**, pp. 177-188.

Kannel, J.W. and Dow, T.A., 1974, The Evolution of Surface Pressure and Temperature Measurement Techniques for Use in the Study of Lubrication in Metal Rolling, ASME, J. Lubr. Techn., **96**, pp. 611-616.

Karagiozis, A.N., 1986, Experimental Determination of the Strip Temperature Distribution During Hot Rolling, M.Sc.Eng. Thesis, UNB, Fredericton, Canada.

Karagiozis, A.N. and Lenard, J.G., 1985, The Effect of Material Properties on the Coefficient of Friction in Cold Rolling, Proc. Eurotrib 85, Lyon, pp. 2-13.

Karagiozis, A.N. and Lenard, J.G., 1987, On Constitutive Relations for High Temperature Behaviour of Steels, Int. J. Mech. Eng. Educ., **15**, pp. 141-149

Karagiozis, A.N. and Lenard, J.G., 1988, Temperature Distribution in a Slab during Hot Rolling, ASME, J. Eng. Mat. Techn., **110**, pp. 17-21.

Klamecki, B.E., 1980, A Thermodynamic Model of Friction, Wear, **63**, pp. 113-120.

Koura, M.M. and Omar, M.A., 1981, The Effect of Surface Parameters on Friction, Wear, **73**, pp. 235-246.

Kraan, D.J., 1981, Metal Forming Oils, Tribology International, pp. 29-32.

Kuhlmann-Wilsdorf, D., 1981, On the Magnitude of the Coefficient of Friction with and without Adhesion, J. Metallkunde, **72**, pp.832-839.

Kumar, A. and Singhal, L.K., 1987, Coupled Finite Element Modelling for Cold Rolling of Austenitic Stainless Steel Strip, Proc. 4th Int. Steel Rolling Conf., Deauville, vol.2, pp. F13.1-F13.7.

Lahoti, G.D. and Altan, T., 1975, Prediction of Temperature Distribution in Axisymmetric Compression and Torsion, ASME, J. Eng. Mat. Techn., **97**, pp. 113-120.

Lahoti, G.D., Shah, S.N. and Altan, T., 1978, Computer- Aided Analysis of the Deformations and Temperatures in Strip Rolling, ASME, J. Eng. Ind., **100**, pp. 159-166.

Lee, H.C. and Kobayashi, S., 1973, New Solution to Rigid-Plastic Deformation Problems Using a Matrix Method, ASME, J. Eng. Ind., **95**, pp. 865-873.

Lee, P.W., Sims, R.B. and Wright, H., 1963, A Method for Predicting Temperatures in Continuous Hot Strip Mills, J. Iron Steel Inst., **201**, pp. 270-279.

Lenard, J.G., 1981, A Theory of Flat Rolling - the Effect of a Variable Coefficient of Friction, Proc. COBEM, Rio de Janeiro, pp. 13-24.

Lenard, J.G., 1985, Development of an Experimental Facility for Single and Multistage, Constant Strain Rate Compression, ASME, J. Eng. Mat. Techn., **107**, pp. 126-131.

Lenard J.G., 1987, Friction in Cold Rolling, Proc. 4th Tribology Conf., Budapest, pp. 119-122

Lenard, J.G., 1989, Friction in Hot/Cold Rolling, Proc. Eurotrib 89, Holmberg, H. and Nieminen, I., (eds), Helsinki, vol.2, pp. 210-215.

Lenard, J.G. and Karagiozis, A.N., 1989, Accuracy of High Temperature, Constant Strain Rate Flow Curves, Factors That Affect the Precision of Mechanical Tests, ASTM STP 1025, Papirno, R. and Weiss, H.C., (eds), ASTM, Philadelphia, pp. 206-216.

Lenard, J.G. and Pietrzyk, M., 1989, The Predictive Capabilities of a Thermal Model of Flat Rolling, Steel Research, **60**, pp. 403-406.

Lenard, J.G., Wang, F. and Nadkarni, G., 1987, The Role of Constitutive Formulation in the Analysis of Hot Rolling, ASME, J. Eng. Mat. Techn., **109**, pp. 343-349.

Li, G. and Kobayashi, S., 1982, Rigid-Plstic Finite-Element Analysis of Plane Strain Rolling, ASME, J.Eng.Ind., **104**, pp. 55-64.

Li, G. and Kobayashi, S., 1984, Analysis of Spread in Rolling by the Rigid-Plastic Finite-Element Method, Numerical Analysis of Forming Processes, Pittman, J.F.T., Zienkiewicz, O.C., Wood, R.D. and Alexander, J.M., (eds), John Wiley & Sons Ltd, pp. 777-786.

Lim, L.-S. and Lenard J.G., 1984, Study of Friction in Cold Strip Rolling, ASME, J. Eng. Mat. Techn., **106**, pp. 139-146.

Liu, C., Hartley, P., Sturgess, C.E.N. and Rowe, G.W., 1985, Elastic-Plastic Finite-Element Modelling of Cold Rolling of Strip, Int. J. Mech. Sci., **27**, pp. 531-541.

Liu, C., Hartley, P., Sturgess, C.E.N. and Rowe, G.W., 1988, Analysis of Stress and Strain Distributions in Slab Rolling Using An Elastic-Plastic Finite-Element Method, Int. J. Num. Meth. Eng., **25**, pp. 55-66.

Luton, M.J., Immarigeon, J.P. and Jonas, J.J., 1974, Constant True Strain Rate Apparatus for Use with Instron Testing Machines, J. Physics, **7**, pp. 862-864.

MacGregor, C.W., Palme, R.B., 1948, Contact Stresses in the Rolling of Metals, J. Appl. Mech., **70**, pp. 297-302.

Maki, T., Akasaka, K. and Tamura, I., 1982, Dynamic Recrystallization Behaviour of Austenite in Several Low and High Alloy Steels, Proc. Int. Conf. TPMA, Pittsburgh, pp. 217-236.

Marion, R.H., 1978, A Short-Time, High-Temperature Testing Facility, J. Testing and Evaluation, **6**, pp. 3-8.

Matsumoto, H., Oh, S.J. and Kobayashi, S., A 1977, Note on the Matrix Method for Rigid-Plastic Analysis of Ring Compression, Proc. 18th MTDR Conf., London, pp. 3-9.

McQueen, H.J. and Jonas, J.J., 1970, Hot Workability Testing Techniques, Proc. Symp. Relation Between Theory and Practice in Metal Forming, Hoffmanner, A.L., (ed.), Plenum Press, NY, pp. 393-428.

McPherson, D.J., 1974, Contributions to the Theory and Practice of Cold Rolling, Met. Trans., **5**, pp. 2479-2498.

Meierhofer, D.J. and Stelson, K.A., 1987, Measurement of the Interfacial Stresses in Rolling Using the Elastic Deformation of the Roll, ASME, J. Eng. Ind., **109**, pp. 362-369.

Michell, J.H., 1900, Some Elementary Distributions of Stress in Three Dimension, Proc. London Math. Soc., **32**, pp. 23-35.

Mokhtar, M.O.M., 1981, Friction: Is it an Intrinsic Property of Metals?, Wear, **72**, pp. 287-293.

Mokhtar, M.O.M., Zaki, M. and Shawki, G.S.A., 1980, Correlation Between the Frictional Behaviour and the Physical Properties of Metals, Wear, **65**, pp. 29-34.

Morawiecki, M., Sadok, L. and Wosiek, E., 1986, Przerobka Plastyczna: Podstawy Teoretyczne, Slask, Katowice.

Mori, K., Osakada, K. and Oda T., 1982, Simulation of Plane-Strain Rolling by the Rigid-Plastic Finite Element Method, Int. J. Mech. Sci., **24**, pp. 519-527.

Mori, K. and Osakada, K., 1984, Simulation of Three Dimensional Rolling Analysis of Forming Processes, Pittman, J.F.T., Zienkiewicz, O.C., Wood, R.D. and Alexander, J.M., (eds), John Wiley & Sons Ltd., pp. 747-756.

Murata, K., Morise, H., Mitsutsuka, M., Naito, H., Komatsu, T. and Shida, S., 1984, Heat Transfer Between Metals in Contact and its Application to Protection of Rolls, Trans. Iron Steel Inst. Japan, **24**, p. B-309.

Murthy, A. and Lenard, A.N., 1982, Statistical Evaluation of Some Hot Rolling Theories, ASME, J. Eng. Mat. Techn., **104**, pp. 47-52.

Muzalevskii, O.G., and Grishkov, A.I., 1959, Methods for the Combined Investigation of Metal Flow and Distribution of Forces in the Zone of Deformation in Rolling, Izv. VUZ, Mashinostroyeniye, **11**, pp. 82-90.

Nadai, A., 1939, The Forces Required for Rolling Steel Strip Under Tension, J. Appl. Mech., **6**, pp. 54-62.

Nicholas, T., 1981, Tensile Testing of Materials at High Strain Rates, Exp. Mechanics, **21**, pp. 177-185.

Orowan, E., 1943, The Calculation of Roll Pressure in Hot and Cold Flat Rolling, Proc. Inst. Mech. Eng., **150**, pp. 140-167.

Owen, D.R.J. and Hinton, E., 1980, Finite Elements in Plasticity: Theory and Practice, Pineridge Press Ltd., Swansea.

Ozisik, M.N., 1985, Heat Transfer, A Basic Approach, McGraw-Hill.

Parker, H. and Jeswiet, J., 1987, Experimental Measurements of Billet Surface Temperature in Rolling, Trans. NAMRC, **15**, pp. 312-315.

Pavlossoglou, J., 1981, Mathematical Model of the Thermal Field in Continuous Hot Rolling of Strip and Simulation of the Process, Arch. Eisenhuttenwes., **52**, pp. 153-158.

Pavlossoglou, J., 1981, Mathematical Model of the Thermal Field of the Rolls and the Strip in Continuous Hot Rolling Including Heat Losses Under Protective Atmosphere and Without Water Spray Cooling, Arch. Eisenhuttenwes., **52**, pp. 275-279.

Pavlov, I.M., 1955, Obrabotka Metallov Davleniyem, Metallurgizdat, Moskva.

Phan-Thien, N., 1981, On an Elastic Theory of Friction, J. Appl. Mech., **48**, pp. 438-439.

Pietrzyk, M., 1982, Analiza metoda elementow skonczonych procesu walcowania blach i tasm, Hutnik, **49**, pp. 213-217.

Pietrzyk, M., 1983a, Walcowanie wyrobow plaskich - modele matematyczne, Metal. Odlew., no.97, (special issue).

Pietrzyk, M., 1983b, Wspolczynnik ksztaltu strefy odksztalcenia jako kryterium oceny procesu walcowania na goraco, Hutnik, **50**, pp. 12-14.

Pietrzyk, M., 1986, The Effect of the Deformation Zone Geometry on Strain Distribution in the Flat Rolling Process, Metal. Odlew., **12**, pp. 409-420.

Pietrzyk, M. and Glowacki, M., 1985, Metoda elementow skonczonych i mozliwosci jej stosowania w procesie plastycznej przerobki metali, Hutnik, **52**, pp. 234-239.

Pietrzyk, M. and Kusiak, J., 1986, Wplyw wspolczynnika ksztaltu strefy odksztalcenia na plyniecie metalu i rozklad naprezen przy walcowaniu wyrobow plaskich, Proc. Conf. PRZEROBKA PLASTYCZNA'86, Raba Nizna, pp. 173-178.

Pietrzyk, M. and Lenard, J.G., 1988, Experimental Substantiation of Modelling Heat Transfer in Hot Flat Rolling, Proc. 25th Nat. Heat Transfer Conf., Houston, Jacobs, H.R., (ed.), HTD-96, vol.3, pp.47-53.

Pietrzyk, M. and Lenard, J.G., 1989a, A Study of Heat Transfer During Flat Rolling, Proc. Conf. NUMIFORM'89, Thompson, E.G., Wood, R.D., Zienkiewicz, O.C. and Samuelsson, A., (eds), Ft.Collins, pp. 343-348.

Pietrzyk, M. and Lenard, J.G., 1989b, Boundary Conditions in Hot/Cold Flat Rolling, Proc. Conf. COMPLAS'89, Barcelona, pp. 947-958.

Pietrzyk, M. and Zakrzewski, J., 1978, Methods to Determine Roll Elastic Flattening, Metal. Odlew., **4**, pp. 107-124.

Pietrzyk, M., Turczyn, S., Nowakowski, A. and Kusiak, J., 1982, Termomechaniczne modelowanie procesu walcowania w zespolach ciaglych, Hutnik, **49**, pp. 105-109.

Pietrzyk, M., Lenard, J.G. and Sousa, A.C.M., 1989, Study of Temperature Distribution in Strips During Cold Rolling, Int. J. Heat and Technol., **7** pp. 12-25.

Plevy, T.A.H., 1979, The Role of Friction in Metal Working with Particular Reference to Energy Saving in Deep Drawing, Wear, **58**, pp. 1-22.

Polukhin, V.P., 1975, Mathematical Simulation and Computer Analysis of Thin-Strip Rolling Mills, Mir Publishers, Moscow.

Poplawski, J.V. and Seccombe, D.A., 1980, Bethlehem's Contribution to the Mathematical Modelling of Cold Rolling in Tandem Mills, Iron Steel Eng., **57**, pp. 47-58.

Preisendanz, H., Schuler, P. and Koschnitzke, K., 1967, Calculation of Temperature Variations at the Surface of Hot Ingots During Contact with Rolls, Arch. Eisenhuttenwes., **38**, pp. 205-213.

Rabinowicz, J.E., 1965, Friction and Wear of Materials, John Wiley & Sons, N.Y.

Rebelo, N. and Kobayashi, S., 1980, A Coupled Analysis of Viscoplastic Deformation and Heat Transfer - II, Int. J. Mech. Sci., **22**, pp. 707-718.

Roberts, W.L., 1974, Friction in the Hot Rolling of Steel Strip, Iron Steel Eng. Year Book, pp. 301-307.

Roberts, W.L., 1978, Cold Rolling of Steel, Marcel Dekker Inc., N.Y.

Roberts, W.L., 1983, Hot Rolling of Steel, Marcel Dekker Inc., N.Y.

Roberts, W.L., 1988, Flat Processing of Steel, Marcel Dekker Inc., N.Y.

Rosenhow, W.M. and Hartnett, J.P., 1973, Handbook of Heat Transfer, McGraw Hill.

Rowe, W.R., 1977, Principles of Industrial Metalworking Processes, Edward Arnold, London.

Roychoudhuri, R. and Lenard, J.G., 1984, A Mathematical Model of Cold Rolling - Experimental Substantiation, Proc. 1st Int. Conf. Techn. Plast., Tokyo, pp. 1138-1145.

Saeed, U. and Lenard, J.G., 1980, A Comparison of Cold Rolling Theories Based on the Equilibrium Approach, ASME, J. Eng. Mat. Techn., **102**, pp. 223-228.

Sankar, J., Hawkins, D. and McQueen, H.J., 1979, Behaviour of Low Carbon and HSLA Steels During Torsion-Simulated Continuous and Interrupted Hot Rolling Practice, Metals Technol., **6**, pp. 325-331.

Sargent, Jr. L.B., 1978, Lubrication Problems in the Rolling of Aluminum, Proc. 1st Int. Conf. Lubrication Challenges in Metalworking, Chicago, pp. 65-70.

Schey, J.A., 1967, Purposes and Attributes of Metalworking Lubricants, Lubric. Eng., **23**, pp. 193-198.

Schey, J.A., 1980, Material Aspects of Friction and Wear in Manufacturing Processes. J. Metals, **33**, pp. 28-33.

Schey, J.A., 1983, Tribology in Metalworking, ASM, Metals Park, Ohio.

Sellars, C.M., 1980, Physical Metallurgy of Hot Working, Proc. Int. Conf. Hot Working and Forming Processes, Sheffield, pp. 3-15.

Sellars, C.M., 1985, Computer Modelling of Hot Working Processes, Mat. Science and Techn., **1**, pp. 325-332.

Semiatin, S.L., Collins, E.W., Wood, V.E. and Altan, T., 1987, Determination of the Interface Heat Transfer Coefficient for Non- Isothermal Bulk- Forming Processes, ASME, J. Eng. Ind., **109**, pp. 49-57.

Seredynski, T., 1973, Prediction of Plate Cooling During Rolling Mill Operation, J. Iron Steel Inst., **211**, pp. 197-203.

Shah, S.N. and Kobayashi, S., A 1977, Theory of Metal Flow in Axisymmetric Piercing and Extrusion, J. Prod. Eng., **1**, pp. 73-103.

Sheppard, T. and Wright, D.S., 1980, Structural and Temperature Variations During Rolling of Aluminium Slabs, Metals Technol., **7**, pp. 274-281.

Shida, S., 1974, Effect of Carbon Content, Temperature and Strain Rate on Compressive Flow Stress of Carbon Steels, Hitachi Res. Lab. Report, pp. 1-9.

Shida, S. and Awazuhara, H., 1973, Rolling Load and Torque in Cold Rolling, Japan Soc. Techn. Plast., **14**, pp. 267-278.

Shimazaki, Y. and Shiojima, T., 1987, Elastic Strain Effects During Steady-State Rolling of Visco-Elastic Slabs, Int. J. Num. Meth. Eng., **24**, pp. 1659-1669.

Siebel, E. and Lueg, W., 1933, Investigation into the Distribution of Pressure at the Surface of the Material in Contact with the Rolls, Mitt. K. W. Inst. Eisenf., **15**, pp. 1-14.

Sigalla, A., 1957, The Cooling of Hot Steel Strip with Water Jets, J. Iron Steel Inst., **186**, pp. 90-93.

Silvonen, A., Malinen, M. and Korhonen, A.S., 1987, A Finite Element Study of Plane Strain Hot Rolling, Scand. J. Metal., **16**, pp. 103-108.

Sims, R.B., 1954, The Calculation of Roll Force and Torque in Hot Rolling Mills, Proc. Inst. Mech. Eng., **168**, pp. 191-200.

Skolyszewski, A., 1986, Application of the Δ Coefficient to the Analysis of the Floating Plug Drawing, Metal. Odlew., **12**, pp. 177-187.

Smelser, R.E. and Thompson, E.G., 1987, Validation of Flow Formulation for Process Modelling, Proc. Symp. on Advances in Inelastic Analysis, ASME Winter Annual Meeting, Boston, Nakazawa, S., Willam, K. and Rebelo, N., (eds), AMD-88, PED-28, pp. 273-282.

Steindl, S.I. and Rice, W.B., 1973, Measurements of Temperature in the Roll Gap, Annals of CIRP, **22**, pp. 89-91.

Stephenson, D.A., 1983, Friction in Cold Strip Rolling, Wear, **92**, 293-311.

Stevens, P.G., Ivens, K.P. and Harper, P.J., 1971, Increasing Work-Roll Life by Improved Roll Cooling Practice, J. Iron Steel Inst., **209**, pp. 1-11.

Stewart, M.J., 1974, Constant True Strain Rate Compression: The Cam-Plastometer, Can. Met. Quarterly, **13**, pp. 503-509.

Suzuki, H., Hashizume, S., Yabuki, Y., Ichihara, Y., Nakajima, S. and Kenmochi, K., 1968, Studies on the Flow Stress of Metals and Alloys, Report of The Inst. of Ind. Science, Univ. of Tokyo, **18**, pp. 1-102.

Tabata, T. and Masaki, S., 1978, Determination of the Coefficient of Friction in Forging of Porous Metals from Ring Compression, Int. J. Mech. Sci., **20**, pp. 505-512.

Tabor, D., 1981, Friction - the Present State of our Understanding, ASME, J. Lubr. Techn., **103**, pp. 169-179.

Theocaris, P.S., Stassinakis, C.A. and Mamalis, A.G., 1983, Roll Pressure Distribution and Coefficient of Friction in Hot Rolling by Caustics, Int. J. Mech. Sci., **25**, pp. 833-844.

Thompson, E.G., 1982, Inclusion of Elastic Strain Rate in the Analysis of Viscoplastic Flow During Rolling, Int. J. Mech. Sci., **24**, pp. 655-659.

Thompson, E.G. and Berman, H.M., 1984, Steady-State Analysis of Elasto-Viscoplastic Flow During Rolling, Numerical Analysis of Forming Processes, Pittman, J.F.T., Zienkiewicz, O.C., Wood, R.D. and Alexander, J.M., (eds), John Wiley & Sons, pp. 269-283.

Thomsen, E.G., Shabaik, A.H. and Sohrabpour, S., 1977, A Modified Tension Test, ASME, J. Eng. Mat. Techn., **99**, pp. 252-256.

Tiitto, K., Fitzsimmons, G. and DeArdo, A.J., 1983, The Effect of Dynamic Precipitation and Recrystallization on the Hot Flow Behaviour of Nb-V Microalloyed Steel, Acta Met., **31**, pp. 1159-1168.

Touloukian, Y.S. and Buyco, E.H., 1970, Thermophysical Properties of Matter, vol.4, Specific Heat - Metallic Elements and Alloys, IFI Plenum, New York - Washington.

Touloukian, Y.S., Kirby, R.K., Taylor, R.E. and Desai, P.D., 1970a, Thermophysical Properties of Matter, vol.12, Thermal Expansion - Metallic Elements and Alloys, IFI Plenum, New York - Washington.

Touloukian, Y.S., Powell, R.W., Ho, C.Y. and Klemens, P.G., 1970b, Thermophysical Properties of Matter, vol.1, Thermal Conductivity - Metallic Elements and Alloys, IFI Plenum, New York - Washington.

Tselikov, A.I., 1939, The Influence of External Friction and Tension on Roll Face Pressure, Metallurg, **6**, pp. 61-76.

Tselikov, A.I. and Grishkov, A.I., 1970, Tieoriya Prokatki, Metallurgiya, Moskva.

Tseng, A.A., 1984, A Numerical Heat Transfer Analysis of Strip Rolling, ASME, J. Heat Trans., **106**, pp. 512-517.

Turczyn, S., 1981, Analiza procesu walcowania blach i tasm stalowych w warunkach adiabatycznych, Ph.D. dissertation, AGH, Krakow.

Turczyn, S. and Nowakowski, A., 1985, Thermal Balance of the Tandem Cold Rolling Process, Metal. Odlew., **11**, pp. 45-62.

Underwood, L.R., 1950, The Rolling of Metals, John Wiley & Sons, N.Y.

Van Rooyen, G.T. and Backofen, W.A., 1957, Friction in Cold Rolling, J. Iron Steel Inst., **186**, pp. 235-244.

Wanheim, T. and Bay, N., 1978, A Model for Friction in Metal Forming Processes, Annals of CIRP, **27**, pp. 189-194.

Watts, A.B. and Ford, H., 1952, An Experimental Investigation of the Yielding of Strip Between Smooth Dies, Proc. Inst. Mech. Eng., **B1**, pp. 448-453.

Weber, K.K., 1973, Mathematical Modelling of Temperature Variations During the Hot Rolling of Plate and Strip, Neue Hutte, **18**, pp. 285-295.

Weinstein, A.S. Matsufuji, A., 1968, Initial Report on Variable Strain Rate Effect on Deformation of Metals, Iron Steel Eng., **45**, pp. 121-136.

Wilcox, J.R. and Honeycombe, R.W.K., 1984, Hot Ductility of Nb and Al Microalloyed Steels, Metals Technol., **11**, pp. 217-225.

Wilmotte, S., Mignon, J., Economopoulos, M. and Thomas, G., 1980, Model of the Evolution of the Temperature of the Strip in the Hot Strip Mill, CRM, **36**, pp. 35-44.

Woo, K.L. and Thomas, T.R., 1979, Contact of Rough Surfaces: A Review of Experimental Work, Wear, **58**, pp. 331-340.

Wright, R.N., 1976, Practical Use of Mechanical Analysis in Wire Drawing, Wire Technology, **4**, pp. 57-61.

Wusatowski, Z., 1969, Fundamentals of Rolling, Pergamon Press, Oxford.

Wusatowski, Z. and Hoderny, B., 1959, Sily i momenty przy walcowaniu na goraco duzymi gniotami miekkiej stali, Arch. Hutnictwa, **4**, pp. 161-171.

Zheleznov, Yu.H., Tsifrinovich, B.A., Lyambakh, R.V., Romashkevich, L.F. and Savichev, G.T., 1968, Variation in Temperature over the Length of a Strip as it Passes through a Continuous Hot Rolling Mill, Stal, **20**, pp. 854-858.

Zienkiewicz, O.C., 1977, The Finite Element Method, McGraw Hill.

Zienkiewicz, O.C., 1981, Finite Element Methods in Thermal Problems, Numerical Methods in Heat Transfer, Lewis, R.W., Morgan, K. and Zienkiewicz, O.C., (eds), John Wiley & Sons, pp. 1-25.

Zienkiewicz, O.C., Onate, E. and Heinrich, J.C., 1981, A General Formulation for Coupled Thermal Flow of Metals Using Finite Elements., Int. J. Num. Meth. Eng., **17**, pp. 1497-1514.

Zmitrowicz, A.A., 1987, Thermodynamical Model of Contact, Friction and Wear: I. Governing Equations, Wear, **114**, pp.135-168.

Zyuzin, W.I., Brovman, M.Ya. and Mielnikov, A.F., 1964, Soprotivleniye deformacii staley pri goriachey prokatkie, Metallurgiya, Moskva.

AUTHOR INDEX

SUBJECT INDEX